Aces and Adventurers

Murray McLeod

Copyright © Murray McLeod 2018

All rights reserved. This book is copyright protected. Apart from any fair dealings for the purpose of private study, criticism, research or review as permitted under the *Copyright Act (Australia),* no part may be reproduced by any process without written permission. Enquiries should be addressed to the publisher.

Original Artwork © by the Author

ISBN: 978-1480167933

A copy of this book can be found in the National Library of Australia.

The author has taken all possible care to give appropriate acknowledgement and seek permissions from all interested parties and welcomes any further correspondence. Enquiries should be addressed to the publisher.

Dedication
To the earlier generation of airmen and women who dared to challenge the astral elements, at times in the face of derision and disbelief

Dedication..3

Introduction..6

Wright Brothers..9
Orville and Wilbur Wright, 'they created wings to life the world'................. 9

Louis Bleriot..13
Louis Bleriot (1872-1936) - Channel conqueror.. 13

Tom Sopwith...17
Tom Sopwith (1888-1989) Sopwith Camel to Harrier Jet.............................. 17

Anthony Fokker...21
Anthony Fokker (1890-1939) The Flying Dutchman.................................... 21

Roland Garros..25
Roland Garros (1888-1918) He put wings on War...................................... 25

Albert Ball 29
Capt. Albert Ball (1896-1917) VC MC DSO & 2 Bars................................... 29

Raoul Lufbury...33
Major Raoul Lufbery (1890-1918) - Ace of the Lafayette 33

Eddie Rickenbacker..38
Capt. Eddie Rickenbacker (1890-1973) Top US Ace.................................. 38

Frank Luke..43
Lt. Frank Luke (1897-1918) Congressional Medal of Honour 43

Bert Hinkler...46
Bert Hinkler (1892-1933) Australia's Lone Eagle....................................... 46

Charles Lindbergh 51
Charles Lindbergh (1902-1974) The Spirit of St. Louis 51

Charles Kingsford Smith..55
Charles Kingsford Smith (1897-1935) The Immortal Smithy..................... 55

Richard Byrd..61.
Cdr. Richard Byrd (1888-1957) Polar Explorer..61

Jimmy Doolittle..65..
Jimmy Doolittle (1896-1993) Air Racer and Tokyo Raider 65

Amelia Earhart..70...
Amelia Earhart (1897-1937) America's best-loved Aviatrix 70

Jean Batten..75...
Jean Batten (1909-1982) New Zealand's greatest aviator 75

Adolph Malan..81
Group Captain Adolph G. Malan DFC & Bar DSO and Bar......................... 81

Pat Pattle...86
Sqdrn.Ldr. M.T. St.JohnPattle DFC & Bar 86

Douglas Bader .... 90.
Wing Commander Douglas Bader DFC & Bar DSO & Bar 90

Clive Caldwell .... 95.
W/Cdr. C.R. Caldwell DFC DSO 95

Pierre Clostermann 99
W/Cdr. Pierre Clostermann (Legion d'Honneur DFC and Bar) 99

Richard Bong 103
Major Richard Bong- Medal of Honour. 103

Chuck Yeager 108
Capt. Chuck Yeager (1923-) Breaker of the Sound Barrier 1947 108

Author appreciation 112

Author Profile   113 ....

Introduction

Over one hundred years have passed since Wilbur and Orville Wright made their tentative aerial forays on those windswept sand dunes at Kitty Hawk. In their wake came a veritable army of hopeful pioneers determined to leave their impression in this exciting new element. Famous names were to emerge, some to advance as major manufacturers, leaving for a decade or so an indelible imprint on the aviation scene. Sopwith, Fokker, De Havilland and Bleriot were such examples to name just a few; innovative individuals who designed and built the machines that bore their name.

Despite their fragility of construction and powered by engines of doubtful reliability these factors were no impediment to the aspiring record-breaker. Former obstacles such as the English Channel and Mediterranean Sea were in turn conquered. Eventually the mighty Atlantic and Pacific Oceans were traversed by air, regardless of the appalling danger and discomfort that were part of the exercise.

In August 1914 the tragedy that was the Great War erupted and with it an urgency to counter and defeat any technical advances the opposing leaders could devise. It eventually became possible to inflict suffering and death on your fellow man in a bewildering array of weaponry; while far above the slough of despond of the battlefield flew the airmen, Knights of the Air in their colourful Nieuport and Albatros scouts. For these aerial jousters it was generally a brief and spectacular career before the odds of survival turned against them.

In spite of this, a select generation emerged, survivors in the cauldron of the skies, names that became synonymous with the fighter ace. For the Allied cause personalities such as Bishop, Fonck or Rickenbacker can evoke admiration decades later. The same applies to those of the Central Powers where Richtofen, Udet or Voss still command respect.

In post-war skies the stage was set for a feast of record-breaking, beginning notably with Alcock and Brown's North Atlantic crossing in 1919. Australian airmen featured strongly during the following decade, with Bert Hinkler's England-Australia solo flight. Equally significant was Charles Kingsford Smith's Pacific crossing from California to Australia in May 1928.

In the Northern hemisphere Charles Lindbergh gained enduring fame with his non-stop New York-Paris epic in 1927. In the process he became the winner of the prestigious Orteig Prize, a contest that attracted a host of entrants and with it the inevitable fatalities. With the emergence of the 1930s aviation was witness to great technical advances; with those fabric-covered, open cockpit biplanes being consigned to the pages of history. In their place came a new generation of all-metal monoplanes, featuring such refinements as radio, retractable landing gear and variable-pitch propellers.

It was still a time for adventurous airmen and women to challenge the elements. Two of the latter deserve special mention, Amelia Earhart and New Zealand's iconic Jean Batten. Since the late 1920s Amelia Earhart had striven to advance the image of female aviators.

Her list of record-breaking flights was impressive indeed and it was particularly tragic that she should lose her life on the closing stages of a round-the-world flight in 1937. Jean Batten for her part gained almost pop-idol fame following a series of spectacular solo flights where she demonstrated navigational and piloting skills equal to the world's best. She came into prominence in 1934 following a record-breaking England-Australia flight and for the ensuing three years she continued on her stellar career. Then as swiftly as she arrived on the scene she left it, for a self-imposed lifetime of seclusion.

In September 1939 Europe was plunged into war and for the Allied cause the old enemy was once again Germany. Front-line aircraft of the opposing powers bore little comparison to their frail ancestors that struggled bravely into Western Front skies of 1914. In parallel with such progress a new breed of airmen had inherited those hard-won lessons of an earlier time.

Famous names were to emerge from the conflict, aces of a new generation. The author has presented a variety of profiles of those characters in caricature mode; whimsical studies yet still retaining an authentic flavour; an international mix of young men and women in the prime of their youth and at the peak of their profession. Five decades of progress are displayed, from the wood and canvas beginnings to the conquering of the sound barrier. An absorbing bio accompanies each art work which makes for an interesting read for young and old alike.

The Wright Brothers

They created wings to lift the World

In modern parlance Wilbur and Orville Wright could well be termed as the 'Dynamic Duo'. They are generally credited with being the first to achieve proper, controlled sustained flight. However the Wright's claim was disputed for many years by their main rival Glenn Curtiss who claimed that Dr. Samuel Langley preceded the Wrights as designer of the first practical aeroplane. Be that as it may photographic and witness evidence tends to support the brothers from Dayton Ohio in their claim to achieving controlled flight.

Wilbur Wright was born 16 April 1867 in Millville Indiana, preceding Orville by two years. From an early age they teamed to develop various businesses and enterprises. Among the more successful were a printing firm and more significantly a bicycle shop. This venture provided sufficient income for them to pursue their abiding passion for aeronautics.

Wilbur gained inspiration from the experiments of German glider pioneer Otto Lilienthal who achieved a degree of fame with a series of man-carrying gliders. Unfortunately during a flight on 9 August 1896 he lost control, crashed and died. Despite this tragedy, Lilienthal's experiments provided the vital spark that kindled Wilbur and Orville's interest in aviation.

They built a glider in 1900 which they tested by mainly flying it as a kite on the end of a rope. It was controlled by a forward elevator and warping wing tips, very similar to modern ailerons and proved quite successful. Other designs followed, each with new features and improvements, and after three and a half years of patient hard work, scientific reasoning and sheer determination they felt it was time to build a powered version.

December 17 1903 found them back at the isolated settlement of Kitty Hawk, North Carolina where they had made their gliding experiments. It was little more than a tiny name on a large map and the only signs of life were a government weather station and a small group of men clustered round the contraption of stick, string and canvas which the Wrights somewhat optimistically called their 'Flying Machine'.

At 10.35am Orville lay down on the lower wing of his aircraft, started its home-made engine, and after a few minutes released a wire which held it onto a wooden track. As it lumbered forward into the 27 knot wind, Wilbur ran alongside, holding a wing tip to balance it. Suddenly it rose into the air, the beginning of the first controlled and sustained flight in history by a powered aeroplane. It was the fulfilment of man's centuries-old dream of flying with the birds.

By modern standards it was not much of a flight; Orville covered only 120 feet; less than the wingspan of many modern airliners. The flight lasted just 12 seconds; but from that modest beginning have come wings to lift the world. Orville made three more flights on 17 December, covering a distance of 852 feet in 59 seconds on the last one. It should have been the story of a lifetime for the American Press, yet only a handful of newspapers found room for a garbled, half-sceptical report of an experiment at Kitty Hawk, North Carolina.

Such a ground-breaking event should have been of great satisfaction for the Wright brothers, but even in America their claim was disputed for an unreasonable period of time.

On two occasions during 1905 the Wrights offered their flying machine to the U.S. Government but traditional conservatism prevented military officials from accepting as true the claims made by the Wrights. In the spring of 1907 however, leading members of the Aero Club of America brought the possibilities of aircraft to President Roosevelt's attention, who had the foresight to order the Board of Ordinance to investigate the claims of the Wright Brothers and to determine the potential of aircraft as a military weapon.

After a series of conferences with Wilbur Wright, the War Department in 1907 called for bids on the purchase of an aircraft. Of forty one bids, three were accepted in February 1908, but only the Wight Flyer fulfilled the requirements. For the acceptance trials which began at Fort Myer, Virginia in August 1908 the Wrights had modified their 1905 model into a two-seater and increased the horsepower to 35hp. However the initial trials were marred by an unfortunate accident in which Wilbur Wright was seriously injured and his passenger Lieutenant Thomas tragically killed.

The trials were resumed in the summer of 1909, and with Orville Wright at the controls the modified Flyer was successful in meeting the requirements. This was enough to win them a $5000 bonus over the contract price of $25000, quite a sizeable amount and a just reward for their earlier endeavours. When the Wright Flyer was accepted on 2 August it became the Army's Aeroplane No1.

Over the ensuing years they were involved in bitter litigation, mainly concerning their attempts to patent the 'Flying Machine'. The main litigant was rival manufacturer Glenn Curtiss and from 1910 until his death from typhoid in 1912 Wilbur took the leading role in the patent struggle. Orville and his sister Katherine tended to blame Curtiss for Wilbur's premature death, which occurred in the wake of his exhausting travels and the stresses of the legal battle.

Orville succeeded to the presidency of the Wright Company upon Wilbur's death, and on his retirement from business he became the elder statesman of aviation, serving on various official boards and committees. Orville died 30 January 1948 having lived from the horse- and- buggy age to the dawn of supersonic flight. Both brothers are buried at the family plot at Woodland Cemetery Dayton Ohio.

Louis Bleriot

Channel conqueror 1872-1936

Louis Bleriot can justifiably be regarded as one of the architects of French aviation, both as a pioneering airman and equally significantly as an aircraft manufacturer. Louis Charles Joseph Bleriot was born 1 July 1872 in the village of Deheries near Cambrai. He studied engineering at the Ecole Centrale Paris and on completion he applied those skills to the design of motor car headlamps. Motoring was in its infancy when Bleriot established an acetylene headlamp manufacturing business and such was the enterprise's success that its creator amassed a modest fortune. This boost to his finances gave him the opportunity to direct his energies to the theories of flight dynamics and aircraft construction.

Between 1903 *(the year the Wright brothers made aerial history)* and 1906 Bleriot collaborated with Gabriel Voisin to form the Bleriot-Voisin Company. A succession of designs emerged from this liaison, all of which proved unsuccessful while some were downright dangerous. By then Bleriot's finances were seriously depleted, prompting him to sever his ties with Voisin and branch out on his own. He concentrated on a monoplane design which was a bold move at a time when multiplanes were the focus of most contemporary manufacturers. His early Bleriot V was far from successful with its propensity for crashing, however 1909 saw the emergence of the Bleriot XI, a machine that would propel its designer into history. It was first displayed at a prestigious aviation exposition in Paris where it drew favourable comments.

At that time Lord Northcliffe, owner of the London Daily Mail newspaper had offered a prize of £1000 for the first successful aeroplane crossing of the English Channel. Already the Channel had been crossed by air, although this was achieved by two intrepid balloonists Blanchard and Jeffries in a hydrogen-filled balloon in January 1785. Given the marginal performance of contemporary aircraft in 1909 this Channel crossing offered a distinct challenge. Confident of his piloting skills and of his aircraft Bleriot became a contender for the Daily Mail prize, however he had two rivals in the contest. Hubert Latham in an Antoinette monoplane would have to be considered as the most serious and already he was well advanced in his preparations. The other was the Comte de Lambert with a Wright biplane and was not taken to be a serious contender however it was Latham who initiated the start of the contest when at 7am 19 July he took off from Sangatte. Bleriot's opportunity seemed to have gone but Latham had barely begun his odyssey when only a few miles over the Channel the Antoinette's engine failed and the machine came down in the sea. Latham was rescued unhurt by the French destroyer which had been following him but the Antoinette was damaged beyond repair. Still determined to make the first Channel crossing and undaunted by this setback, within three days Latham had obtained another machine.

For Bleriot, his moment of destiny was at hand and despite the discomfort of a badly- burned foot, suffered in one of his numerous flying accidents he was ready to leave. It was 4am 25 July 1909 and in ideal weather conditions the little monoplane staggered into the air from a field at Les Baraques near Calais, its destination Dover. Incredibly its pilot carried no compass and for the early stage of the flight he steered by the smoke from the destroyer *Escopette* which the French government had provided as an escort and to observe his progress.

With an average speed of 40mph Bleriot soon overtook the vessel and with the onset of rough weather he lost sight of any landmarks. He made no attempt to steer and for the final twenty minutes the aeroplane took its own course, until to his immense relief the English coastline appeared. It was St. Margaret's Bay and not Dover that he saw, the wind having carried him slightly off course; nevertheless it was a remarkable achievement considering the lack of navigational aids. A strong wind had sprung up and Bleriot was obliged to land in a meadow behind Dover Castle. Today the outline of an aircraft laid out in granite blocks depicts the exact landing spot on the cliffs above Dover. Bleriot had flown 22 miles from Les Baraques to Dover in a time of 37 minutes, with his machine and also its pilot suffering some damage in the ensuing heavy landing. Surprisingly, considering this historic event when Bleriot stepped out of his machine there was no one to witness his arrival. Several minutes passed before a policeman arrived, then came a French journalist who was to have marked the predicted landing place and in turn he was followed by a small crowd of people. They soon dispersed and one can only assume they were unimpressed. It might be imagined that the feat would be the talk of Dover, almost as soon as Bleriot arrived but this was not so. Later the feat was widely acclaimed and just as soon forgotten.

Bleriot's crossing marked a welcome change in his personal fortunes, for apart from the £1000 offered by Lord Northcliffe he won also a £2000 prize from France. Altogether he made £4000 out of his flight and with these funds he opened a flying school at Pau, France and began to design and build aircraft which would become world famous. Between 1910 and 1914 by way of contracts with the French government Bleriot built 800 machines of 40 different types, thus making him the world's largest aeroplane manufacturer.

With the outbreak of war in 1914 Bleriot's machines were initially in great demand but as the conflict progressed they were subsequently not used in large numbers. Bleriot had concentrated his efforts on the Deperdussin works which he had bought, naming it the *Societe pour I'Aviatonetses Derives* (SPAD). Among the most famous of its products was the excellent SPAD scout, which in various types was flown by British, French and American pilots during the war. At the end of hostilities Bleriot's works were producing 18 aircraft a day. Following the armistice Bleriot personally supervised the design and construction of a wide variety of aircraft, including high-speed machines and large four- engine flying boats.

In 1927 Bleriot was one of those present at le Bourget field to welcome Charles Lindbergh following his historic non-stop New York to Paris flight. It was a significant meeting, for these two men, separated in age by thirty years had each made an historic flight over famous bodies of water. Still active in the aviation industry Bleriot died of a heart attack in Paris on 22 August 1936 and was later buried at Versailles. On 25 July 2009, the centenary of the original Channel crossing, a French pilot Edmond Salis re-created the event in an exact replica of the original Bleriot XI, taking off from Bleriot Beach and landing in Kent at the Duke of York's Royal Military School. This was a significant achievement by Salis and a tangible reminder of an intrepid journey a century ago.

Tom Sopwith

Sopwith Camel to Harrier Jet

T.O.M. Sopwith is rightly remembered as one of the luminaries of British aviation design and manufacture. In a lifetime that spanned a remarkable 101 years he outlived all his contemporaries and was witness to an evolution in aviation from those hesitant Edwardian beginnings to the Harrier jump-jet of contemporary times. He also led the companies which designed the highest-scoring British fighter of World War 1, the Sopwith Camel and the highest-scoring fighter of the Battle of Britain, the Hawker Hurricane.

Thomas Octave Murdoch Sopwith was born on 18 January 1888, in London, the eighth child (hence Octave) and first son of Thomas Sopwith, civil engineer. When Sopwith was born, in those far-off days powered aeroplane flight was still fifteen years away. As a young man, sailing was one of Sopwith's abiding passions, an interest shared with motor cars, power boats and ballooning. In fact his interest in ballooning encouraged him to become involved in powered flight.

His first experience was at Brooklands in the summer of 1910 where a French pilot Gustav Blondeau was offering passenger flights in a Maurice Farman. The cost was £5, a considerable sum of money at the time but by all accounts Sopwith was fairly well placed in financial terms. Exhilarated by the Brooklands flight Sopwith was determined to become a pilot himself. This was achieved in the autumn of 1910 when he purchased a Wright-Manning biplane. He assembled it at Brooklands and began a series of self-taught exercises in taxying and take-offs. These resulted in a number of crashes, fortunately without serious injury and on 22 November 1910 Sopwith was awarded Aviation Certificate No31. The prestigious No.1 Certificate was held by an associate of Sopwith, J.T.C. Moore-Brabazon who made headline news when he took aloft a pig secured in a wicker basket thus de-bunking the long-held cynical suggestion that 'pigs might fly'.

In December 1910 Sopwith made the decision to compete for a prize of £4000 offered by the Baron de Forest for the longest flight from England to the Continent in a British-built machine. Popular opinion was that the most probable risk of engine failure occurred during take-off and climb. Sopwith took the precaution of leaving from Newchurch (half an hour inland) rather than Dover. A cross-Channel flight was largely a step into the unknown and Sopwith later admitted to an intense loneliness until the relative haven of the French coast was reached. He eventually landed at Beaumont, having flown 169 miles to claim the substantial prize.

Four days later a friend of Sopwith's, Cecil Grace attempted to better Sopwith's cross-Channel flight. Bad weather at Calais turned him back and near Dover he encountered heavy fog and was last heard heading out over the North Sea, never to be seen again. Such events were saddening features of the period, in particular the death of Sopwith's close friend S.C. Rolls at a Bournemouth meeting. Charles Rolls was in partnership with the talented engineer Henry Royce. Their reputation for fine motor cars was already established and with their ongoing range of prestigious aero engines it forged; a tradition that endures to this very day.

In 1911 Sopwith made a tour of The United States where he conducted a series of exhibition flights. In the process he won an amount of prize money, despite the occasional crash landings that were hazards to be endured by those pioneers. Returning to England in 1912 he formed the Sopwith Aviation Company, starting at Brooklands with a flying school. The company's first product was known as the Sopwith-Weight Tractor, which attracted Admiralty interest and was purchased for £900 for naval use. By the time the Kaiser's war started in August 1914 Sopwith had assembled a worthy team of designers, most notable was Aussie ex-pat Harry Hawker whom Sopwith had taught to fly and went on to become the firm's chief test pilot.

Throughout that period Sopwith Aviation was involved in the design of at least 80 types although only a percentage of these entered production. Those that did went on to establish enviable reputations; the delightful Pup and 2-seat Strutter were early examples. These were followed into service by the iconic Camel which gained notoriety as being tricky to fly. Nevertheless in capable hands it was a formidable fighting machine and with a total of 1300 victories it was the top-scoring British scout of World War 1. Sopwith had produced 6000 examples of the F1 and 2F1 Camel which by late 1918 would have been superseded by the Snipe which continued in front line RAF service until 1924.

At the time of the armistice Sopwith Aviation had grown into an impressive business, having produced 16,000 aircraft and employing 5,000 workers plus thousands more through subcontractors. But with the end of hostilities demand for military equipment vanished overnight making it impossible to do anything but to cease operations and go into voluntary liquidation. Sopwith kept his design team together and with new capital he hired an area of the old Kingston works from the liquidators. The new firm would become the Hawker Company with Sopwith as its chairman.

In 1923 the firm was joined by Sydney Camm whose future designs; epitomised by the Fury, Hurricane, Tempest and Harrier established Hawker Aviation as arguably the leader in British fighter development. In 1978 Hawker aerospace interests in Britain were absorbed into British Aerospace. This continued with the engineering group Hawker Siddeley and in Hawker De Havilland of Australia. Throughout this period Sir Thomas Sopwith (knighted in 1953) was three times president of the SBAC, the supreme body of Britain's aerospace industry. He was also president of the Hawker Siddeley group until he retired at the age of 90 in 1978 while still maintaining an interest in aviation, alert and discerning until his death in 1989 at the grand old age of 101.

Anthony Fokker

The Flying Dutchman (1890-1940)

Anthony Fokker merits inclusion as one of the pioneer designers and manufacturers in aviation's formative years. He was born in 1890 in Blitar, a settlement in the Dutch East Indies (now Indonesia). His father Hermann Fokker owned a Dutch coffee plantation and when Anthony was four the family returned to the Netherlands to settle in Harlem. Although not of a studious nature and failing to complete his high school education, Anthony showed an early interest in mechanics. Model trains and steam engines were interests that eventually led him to experiment with model planes.

In 1908 Wilbur Wright was making exhibition flights in France with one of his biplanes, which inspired the current French designers to even greater efforts. Anthony Fokker was equally impressed with this new concept and in 1910, aged 20 he was sent by his father to Germany as a mechanic at Bingen Technical School. However his abiding interest was in aviation which resulted in him transferring to an aeronautical college at Mainz.

That same year Fokker built his first aircraft; it was named De Spin (The Spider) but unfortunately it was destroyed in a crash when his business partner flew it into a tree. A second Spider was built, in which Fokker gained his pilot licence and added to his fame with exhibition flights in his home country. In 1912 Fokker moved to Johannisthal near Berlin where he founded his own company Aeroplanblau. Over the following years the innovative Fokker constructed a variety of aircraft and shortly he relocated his factory to Schwerin.

At the outbreak of war in August 1914 the German government took control of the factory, with Fokker retained as director. A wide range of designs was produced, some quite outstanding for their time and others less so. Their E series Eindecker gained notoriety over the first 18 months of the conflict with its overwhelming supremacy over current RFC types, thus creating the so-called Fokker 'scourge'. The factor in this supremacy was the installation of synchronised machine gun to the nimble Eindecker, a feature that was denied to the hapless RFC crews at that time. Several German aces, notably Oswald Boelcke and Max Immelman gained their initial fame flying the Eindecker. Fokker is often credited with having invented the synchronisation device which enabled the gun to fire through the spinning propeller. His role was certainly significant but there were a number of prior developments before the desired result was achieved for which Fokker is generally credited.

Further development of the Eindecker continued, with the generally unsuccessful fitting of firstly two and later three machine guns. However by January 1916 it began to encounter sterner opposition in the form of the DH2, Nieuport and FE2b. It was an FE2b that finally ended Immelman's glittering career when he was shot down by Lt. McCubbin on the evening of 18 June 1916, although German sources claim he crashed due to structural failure of the airframe.

Fokker then concentrated on the biplane 'D' series fighters, from the D1 to the DV, however these were generally mediocre in performance when compared to the ascendant Albatros types in front line service. In late 1917 Fokker introduced a type that gained more notoriety than any other World War 1 aircraft.

This was the DR1 triplane that gained everlasting fame in the hands of Manfred Richtofen, who initially rejected the DR1 but later he became a devotee of the type. Other German aces, notably Werner Voss and Lothar Richtofen were exponents of the DR1, and it was in a Dr1 that Voss fought his epic solo duel against overwhelming odds on 23 September 1917

Perhaps the DRI's reputation was oversubscribed, for its front line service was comparatively brief, to be supplanted by the Fokker DVII, which was generally regarded as the finest fighter of the war. A significant feature of the November Armistice terms was that all the DVIIs be handed over intact to the Allies. The Treaty of Versailles also forbad Germany to build any aircraft or engines; this outcome prompted Fokker to return to the Netherlands to start a new company, later to become the Fokker Aircraft Company. Despite the strict disarmament conditions of the Treaty, Fokker managed to smuggle an entire train load of DVII and CI planes and spares across the German/Dutch border; this initial stock enabled him to quickly set up shop in his homeland.

From then on his focus shifted from military to civil types, typified by the successful FVII series, which during the 1920s in various forms advanced the cause of civil aviation possibly more than any other type. In 1922 Fokker moved to The United States and later became an American citizen. He established the North American branch of his company, The Atlantic Aircraft Corporation, which in 1931 he sold to General Motors where it became their General Aviation Division.

Fokker died in New York in 1939 from meningitis, having been ill for three weeks; unfortunately the anti-biotics that could have saved him were simply not available in those pre-war years. Anthony Fokker was 49. In 1940 his ashes were brought to Westerveld Cemetery in Drichius where they were buried in the family grave.

Roland Garros

Roland Garros (1888-1918) He put wings on war

Roland Garros can well be regarded as one of the pioneers of aviation in the period leading up to World War I. He was born at Saint-Denis in October 1888 and following a tertiary education at technical institutions Garros directed his energies towards an aviation career. By all accounts he was a young man of independent means, enabling him to indulge in expensive motor cars and more significantly the embryonic aircraft of that era. His small stature was well-suited to the low-powered aircraft of 1909 and flying the iconic Bleriot XI the intrepid Garros gained acclaim at European air races. In 1913 he changed to a type with which he would forever be associated, the Morane-Saulnier parasol. Perhaps his greatest achievement in the diminutive Morane was for making the first non-stop Mediterranean crossing from the south of France to Tunisia.

In July 1914 Garros was in Berlin making exhibition flights in his Morane Parasol. By early August it was obvious that war with France and Germany was inevitable and the wily Garros had no intention of presenting himself and his aircraft as prisoners of war to Germany. His dilemma was how to make a getaway from under the noses of his German hosts who shortly would be his captors. If the account of his daring night escape from a Berlin airfield was factual it belongs with the classics but Garros was seemingly undaunted by the odds.

After managing to evade his captors Garros made his way to the airfield where his Morane was stored. By rights it should have been under armed guard but to his relief it was unattended. He was then faced with the problem of starting it single-handed, a procedure normally involving several assistants. Undismayed Garros pointed the Morane at an area clear of obstacles, swung the propeller and before his machine bolted into the darkness he hauled himself onto the wing and clambered into the cockpit. Somehow he coaxed the Morane into the air and navigating by the stars he carried out an amazing escape. Back on French soil he wasted no time in joining the French Air Service where he flew reconnaissance missions with NS26. Disenchanted with this role the aggressive Garros saw it as his duty to restrict the activities of his German counterparts with whatever means it took.

Early attempts to mount a machine gun on a tripod to fire above the propeller blades had proved unrewarding, should the gun jam which was fairly common it was impossible to clear the stoppage in flight. On a visit to the Saulnier works in December 1914 Garros became enthusiastic about the firm's approach to the problem with the installation of metal deflectors on the engine's propeller blades. At that time the Allies had no access to an interrupter gear to fire through the airscrew and Saulnier's solution was extremely basic and opposed to engineering principles. It was also quite daunting for the aspiring airman, however Garros reasoned that although a fair number of bullets would strike his propeller enough would pass through intact to down an enemy aircraft.

Perhaps it was significant that Garros should launch his device on April Fool's Day 1915. His initial encounter was with four Albatros 2-seaters over the French lines north of Paris and at first the Germans paid scant attention to his Morane as it bore down on them.

Their disdain turned to horror when they became aware of flames shooting between his propeller blades and one of their comrades fell away out of control. So close did Garros fly to the doomed Albatros that his companions mistakenly presumed the Morane had clipped the upper wing of the Albatros but in that brief engagement Garros had changed forever the tenor of aerial combat

For almost three weeks he ranged unopposed in his particular sector, a period when 3 German 2-seaters fell to his gun. But time was running out for the intrepid Garros when on 18 April the engine of his long-suffering Morane finally succumbed to the stresses being imposed on it. Garros made an emergency landing behind enemy lines and before he could destroy his machine he became a prisoner of war for the second time. The identity of its pilot was soon revealed and once again Garros and his plane were on their way to Berlin and on this occasion over an extended period for its celebrated pilot.

According to legend Anthony Fokker was summoned to make an appraisal of the Garros device. He summarily dismissed it for its crudity and in any event the Fokker works were already well advanced with their own interrupter gear. Fokker was shortly able to demonstrate the synchronised gun installed in an Eindecker scout. In no time several examples became operational and it was the Allies' turn to be driven from the skies. The so-called Fokker 'scourge' had begun, expounded in no small measure by two emerging aces, the redoubtable Oswald Boelcke and Max Immelman.

Garros' career had been marking time during his sojourn as a P.O.W. but in February 1918 after a series of unsuccessful escape attempts he made a bold getaway. His guards were not averse to bribery and on a pitch-black night he was flown to freedom from an airfield near Cologne. He was soon in the air again, this time flying a Spad with Esc.26. Garros was credited with one more victory, thus taking his victory tally to four and one short of gaining 'ace' status. But in that hostile environment the law of averages was against him and in October 1918 he fell to the guns of a Fokker DVII. This was a particularly tragic outcome to lose his life only weeks before the armistice and one day before his 30th birthday. It was a fitting tribute that his name be commemorated with the Roland Garros tennis stadium in cosmopolitan Paris.

Albert Ball VC.

Capt. Albert Ball (1896-1917) VC MC DSO & 2 Bars

Pioneering RFC airmen faced conditions without precedent and accepted willingly the challenge to perform their duties in an unforgiving element. This was achieved in machines of doubtful reliability and construction and without the security of parachutes. It was a matter of creating the basic rules and techniques of air fighting. Among these early fighting pilots there were many who earned a distinguished niche in RFC history. Inevitably it was the fighter ace that became the focus of media attention rather than the bomber or reconnaissance pilot. The latter were obliged to endure the perils of their dangerous vocation while rarely being the object of the adulation bestowed on their counterparts.

Many of these 'Knights of the Air' attained greatness in what was generally a brief and turbulent career. A shining example was Nottingham's favourite son Albert Ball, who for a time blazed a fiery path over the Western Front. He was born in August 1896 and from an early age was involved in hobbies that were of a practical nature. This was borne out while still a teenager when he set up his own brass founding and electrical business.

At the outbreak of war Ball enlisted in the Sherwood Foresters and in October he received his commission, although he did not go on active service immediately. However his destiny was changed dramatically, when following a visit to Hendon Aerodrome Ball became so enthusiastic about flying that he determined to become a pilot himself. On gaining his certificate he was posted to the R.F.C. and in February 1916 he joined No.13 Squadron.

His new squadron was a reconnaissance unit based in France, operating with the obsolescent BE2C 2-seater. Ball had little opportunity for combat with No.13, although in an encounter which was inconclusive he did make an attack on an enemy scout. In May 1916 he transferred to No. 11 Squadron, equipped with Newport and Bristol Scouts. Ball developed an immediate rapport with both types and quickly established himself as an aggressive fighter pilot. After earlier unconfirmed combats, by July he was credited with the destruction of a balloon and also a Roland CII.

The Battle of the Somme was launched on 1 July 1916 and to Ball's disgust he was posted to a reconnaissance unit, No.8. Over the ensuing weeks he flew the outdated BE2d on photo/reconnaissance missions over a bitterly contested battleground. During his tenure with No.8 he attacked an enemy balloon, forcing its observer to take to his parachute. However his period with No.8 was only for a short spell and to his relief on 10 August he returned to No.11. In the course of a solo patrol Ball attacked a formation of 5 enemy machines and in the ensuing combat he forced down four of them. On 21 August he attacked 7 Roland CIIs and once again ignoring the odds he shot down three of them. For a person of such an individualistic nature, Albert Ball was in his element on these roving commissions. During September, undeterred by the odds facing him, more enemy machines fell to his guns and for these and earlier exploits the buccaneering Albert Ball was awarded the DSO and Bar.

Ball was seemingly unperturbed by the odds against him in his combats although in retrospect his methods bordered on suicidal. He was a somewhat introverted character and preferred to isolate himself on his off-duty periods. Much of his time was spent in a self-contained hut set apart from the regular squadron accommodation. There he was able to tend a small vegetable garden beside his front door. The riotous mess parties were events from which he preferred to disassociate himself.

By early October 1916 the General Staff considered Ball had done more than enough operationally and he was sent back to England on instructional duties. At the time he had destroyed 10 enemy aircraft and forced down a further 20. He soon tired of instructing pupils and applied for a return to France. He was forced to wait until 7 April 1917 before his transfer was approved and the move when it came was to No.56 squadron as a flight commander. It was already a famous unit, and the first to be equipped with the SE5 scout. Ball took some time to become reconciled to the SE5 and for a time his preference was to mostly operate with his faithful Nieuport scout.

On the evening of 7 May 1917 Ball was leading a patrol of eleven SE5s when they encountered aircraft from Jasta II. One of the Jasta pilots was Lothar von Richtofen, younger brother of the celebrated Manfred and during the subsequent combat Ball was seen to give chase to Lothar's Albatros DIII. Lothar was forced down unhurt with a damaged fuel tank and is thought to have observed Ball's SE5 disappear in the heavy cloud conditions. A German officer on the ground then watched the SE5 emerge from the cloud, inverted and with its propeller stopped, crashing close to a ruined farmhouse. A young Frenchwoman dragged Ball from the wreck and there in her arms he gasped out his last precious moments.

It would appear that he became disoriented in the heavy cloud and while in an inverted situation his SE5, an early example would not have maintained power, leaving Ball with no hope of retrieval. On 9 May Ball was buried by the Germans and accorded full military honours. For conspicuous gallantry between 25 April and 6 May Albert Ball was awarded a posthumous Victoria Cross. This was on 8 June 1917 and his impressive array of decorations comprise the Victoria Cross, DSO and 2 Bars, Military Cross, 1914-15 Star, British War Medal, Legion d'Honneur and Order of St. George. His final victory tally was 43 aircraft and one balloon.

Raoul Lufbury

Major Raoul Lufbury (1890-1918) Ace of the Lafayette

With seventeen confirmed victories Raoul Gervais Lufbery merits inclusion in the list of World War 1 aces; this tally ranking him third of the American aces behind Eddie Rickenbacker and Frank Luke. He was born in 1885 at Clermont-Ferrand, of French parents and on the death of his mother in 1886 his father placed Raoul in the temporary care of a family in the Auvergne Mountains. In 1890 his father re-married and the following year the family immigrated to America where they settled in Wallingford Connecticut. Once again Lufbery senior suffered bereavement when his second wife died in 1901 and no doubt these events unsettled Raoul. The youngster was always of a restless nature and after a period of employment with his father and bored with attending school he left home and from there he began a career of world travel and adventure.

Over the following four years he journeyed through France, North Africa and the Balkans, working at any task to finance these adventures. On his return to America in 1906 to visit his father, who unaware of his son's impending arrival had just left on a world trip to gather rare stamps for his collection. It was a bitter-sweet situation for young Raoul for the pair was destined never to meet again.

After a two years sojourn in America Raoul embarked on a second venture, this time to Cuba and from there to cosmopolitan New Orleans where he worked for a time in a bakery. In 1908 he enlisted in the US Army and spent two years in the Philippines where he distinguished himself as a rifle marksman. With the finish of his military tenure Lufbery sought new adventures in Japan and later in China where he found employment in the Chinese Customs service.

After a period spent in India, Lufbery then travelled to Saigon, where he made the acquaintance of another itinerant Frenchman, the aviator Marc Pourp. The intrepid Pourp was making a series of exhibition flights in a Bleriot monoplane and despite Lufbery's inexperience he engaged him as a mechanic and general handyman. Lufbery soon adapted to the theory of flight and mechanics and for the next year or so they toured China and the Far East. The summer of 1914 saw them in France where Pourp was to take delivery of a new Morane Parasol for a proposed tour of the Orient.

With the outbreak of war these plans were cancelled and Pourp immediately enlisted in the French Air Service, being sent to Escadrille N.23. Disappointed with the break-up of their peacetime flying operation Lufbery was eager to join the service himself but as an American citizen he was denied entry. Lufbery circumvented the problem by enlisting in the French Foreign Legion and with Pourp's influence he was able to join N23 as a mechanic. However this unique partnership was destined to last only a few months, when in December 1914 Pourp was killed in action.

No doubt anxious to avenge Pourp's death Lufbery applied for pilot training and this request was granted. Following primary training in the ubiquitous Maurice Farman he received his brevet in July 1915 which saw Lufbery being posted to VB106. This was a bombing and reconnaissance unit, operating with the clumsy Voisin and Caudron types where for a period of six months he participated in almost a daily routine of bombing missions.

The impetuous Lufbery became disenchanted with the monotony of this routine and yearned for the opportunity to become a fighter pilot. Eventually he was transferred to a fighter school at Le Plessis-Belleville for training as a *pilote-de-chasse.* Initially he had difficulty in coming to terms with the agile Nieuport, due to the heavy-handed technique which he developed in his time with the bombing squadron. At times his instructors despaired of Lufbery applying the necessary finesse to his piloting but with application and perseverance this was eventually achieved.

On completion of training he was sent to the Escadrille Lafayette N124 which was an appropriate posting, for its personnel were American soldiers of fortune like the irrepressible Lufbery. He brought down his first victim on

30 July 1916, followed five days later by another, and by 12 October Lufbery had achieved ace status, and as was the French custom was given a citation after each subsequent victory. For these and other exploits Lufbery was awarded many decorations, including the *Croix de Guerre (with ten palms), Medal Militaire* and named a *Chevalier of the Legion of Honour.* He was also the first American to receive from Britain the Military Cross which was bestowed in June 1917, a time when he claimed his tenth victim and was promoted to First Lieutenant. Within several months his tally had risen to seventeen.

Following America's entry into the war Lufbery was commissioned as a major in the US Air Service although he remained with N124 until January 1918 and then transferred to Issoudin in an administrative post. Despite his boisterous personality Lufbery's health was not always on a par and increasingly he was confined to hospital with severe bouts of rheumatism, he still continued to fly as patrol leader whenever the opportunity arose.

In late January he was appointed C.O. of the recently-formed 95th Aero Squadron and on 19 March the unit made the first American sorties over enemy lines. The three-man patrol comprised Major Lufbery, Lt. Campbell and Lt. Rickenbacker, all of whom would achieve fame in the months that followed. In April Lufbery was transferred to the 94th Aero Squadron as commanding officer but his tenure there was destined to be tragically brief.

At 10am on Sunday 9 May an alert was sounded of an approaching enemy aircraft. The intruder, an Albatros 2-seater was engaged by anti-aircraft which appeared to have scored a hit, however after entering a spin it recovered and headed back to its own lines. At that point the impetuous Lufbery, eager to intercept the intruder took off in another pilot's Nieuport 28, his own aircraft being temporary unserviceable.

Shortly he was in a position to attack and in his first pass which was inconclusive he turned away, possibly to clear a jammed machine gun.

Returning once again he attacked from the rear but his burst of fire seemingly had no effect, when to the horror of the spectators on the airfield his Nieuport was suddenly enveloped in flames. Side-slipping his aircraft violently to direct the flames away from the cockpit Lufbery was seen to climb out on the port wing and in a desperate decision that he might land in the Toul Canal he jumped.

Lufbery's hopes were futile; for he smashed onto a picket fence bordering a cottage in the village of Maron where he gasped out his last precious moments in the arms of a peasant woman. When Lufbery's squadron mates arrived at the scene they discovered his body covered with flowers in a final tribute by the village folk who had witnessed his fall. In a touch of irony the ill-omened Nieuport continued to fly with its flames extinguished, until it crashed and burned in total destruction. The rogue Albatros was later shot down and on inspection Lufbery's intended victim was found to be armour-plated. He was buried the following day with full military honours to the rear of the American lines, the adventurous Lufbery had finally returned to the soil of the country of his birth.

Eddie Rickenbacker

Top-scoring U.S. Ace (1890-1973)

Edward Vernon Rickenbacker achieved fame as America's top-ranking ace of World War 1. It was one facet in a career that encompassed automobile manufacturing, motor racing and a host of corporate activities. He was born in Columbus Ohio in 1890, the son of William and Elizabeth Reichenbacher. The Teutonic origins of the family name caused him to change it to an Americanised version as 'Rickenbacker'. Following America's entry into the war he was hounded by their Secret Service and at one stage Scotland Yard actually arrested him; such was the anti-German hysteria that prevailed.

Following the death of his father in 1902 the young Rickenbacker was not afforded the privilege of a secondary education. Instead he was obliged to leave primary school at twelve years of age and find employment to support himself and his widowed mother. Initially he worked at a variety of firms, notably at a garage repairing motor cars where he showed an early aptitude for mechanics. At the same time he advanced his experience through correspondence courses in engineering and draughtsmanship. Rickenbacker's success in mechanics and sales led him to motor racing where he became a consistent winner on American tracks. Over the following six years he became one of the leading U.S. racers, competing in the Indianapolis 500 classic and also creating a world's speed record of 134 mph at Daytona Beach. At the peak of his career he was reputed to be earning $40,000 annually, a fabulous sum in those days.

In April 1917 following America's entry into the war any racing activities were abruptly terminated, prompting Rickenbacker to apply for the pilot's course with the U.S Army Air Service. To his dismay he was initially rejected due to his age (27 years) and lack of a college education. As an alternative he enlisted in the Army and because of his fame as a race driver he was assigned as a personal driver to General Pershing, G.O.C. of the American Expeditionary Force. Although the posting was not entirely to his liking it did offer opportunities to advance his aviation prospects through his contacts with General Billy Mitchell, chief of the Air Service, whose influence saw Rickenbacker being accepted for pilot training.

Following an intensive course at the Aviation Instruction Centre Rickenbacker qualified as a pilot in the remarkable time of 17 days. However his engineering background was so regarded that rather than being posted to a fighter unit he was assigned to 3rd Aviation Instruction Centre as chief engineering officer. Impatient to be involved in aerial combat it took many requests for Rickenbacker to be released from his engineering situation, until finally he became a pilot with the newly formed 94th Aero Squadron. Their initial equipment was the rotary engine Nieuport 28, and although it represented the peak of the Nieuport scout's development it was powered by the thoroughly unreliable Gnome Monosoupape engine and also earned an unwanted reputation for shedding its upper wing fabric in any violent manoeuvre. Rickenbacker's C.O. and mentor was the mercurial soldier of fortune Raoul Lufbury who on March 19 led the first American patrol over enemy lines. The three pilots on that significant occasion were Major Lufbury, Lt.Campbell and Lt.Rickenbacker. His first victory was on 29 April and by 30 May he achieved 'ace' status with his fifth.

At age 27 Rickenbacker was older than the average pilot, which was reflected in his judgment and maturity in combat flying. The period between June and August 1918 saw Rickenbacker recovering from a mastoid operation and it is reasonable to assume that but for that setback he may well have doubled his score. It was September before he returned to duty and by the end of the month he achieved 6 more victories. During that period he was appointed C.O. of the 94th Aero Squadron, holding that position until the Armistice. The squadron had recently exchanged its Nieuports for the more amenable Spad XIII, which Rickenbacker used to good effect to add a further 14 to his score during October. At the end of hostilities on 11

November his score stood at 26, making him America's 'Ace of aces' and also a recipient of the Congressional Medal of Honour.

Rickenbacker returned home to a hero's welcome and was bombarded with lucrative offers from movie and advertising companies. These approaches were summarily dismissed and he involved himself in a host of corporate projects. These were associated with aviation and automobiles, and one that deserves mention was the manufacture of one of the finest vehicles of the time. Naturally enough it was named the 'Rickenbacker', and it incorporated features that later became standard throughout the industry. But the project was not sufficiently funded to withstand the fluctuations of Wall Street and like so many other makes it subsequently failed. In 1927 he acquired an interest in the Indianapolis Speedway and served as president of the company until he disposed of his interests in 1945.

In 1926 Rickenbacker began an involvement with commercial aviation, founding Florida Airways which was shortly sold to an emerging Pan American Airways. After serving as vice president with General Aviation Corporation he joined North American Aviation as vice president and then moved to its subsidiary Eastern Air Transport which was eventually re-organised as Eastern Air Lines. His tenure with Eastern began in 1934, a period when the Army Air Corps were poised to take over the airmail contracts. Rickenbacker was opposed to such a move and to demonstrate the civil airline's capacity to continue the contracts he flew the sole Douglas DC1 ever built in a coast to coast record of just over 13 hours. Under his leadership Eastern advanced its place in American civil aviation, showing its first profit since its inauguration. The ageing fleet was replaced by the new 14-passenger DC2 and by 1938 Rickenbacker and several associates were in a position to purchase the re-born airline with Rickenbacker elected as president and general manager.

By 1941 the airline was serving 40 cities with their fleet of iconic DC3s but with the onset of war it was faced with major changes. Eastern was obliged to hand over half of its fleet to the armed services, while many of its pilots left to join the Army Air Corps. Once again Rickenbacker volunteered to serve his country; this time as a non-military observer for Secretary of War Henry Stimson. This tenure involved a great deal of travel and a measure of Rickenbacker's durability was displayed during that period when on such a flight his aircraft was forced to ditch in the Pacific Ocean. In an epic demonstration of survival he and his companions existed on limited rations for 21 days until their rescue.

This experience convinced Rickenbacker as to the absolute necessity of adequate survival gear aboard military aircraft, with such measures being installed as standard equipment.

In a post-war situation Rickenbacker focused once more on Eastern Airlines, expanding routes and updating the fleet with Lockheed Constellations and Douglas DC4s. In 1953 after 19 years of service he moved up to be chairman of the board but found it difficult to give control to the new president, until in 1963 he finally retired to a ranch in Texas with his wife Adelaide, however the remoteness of the property did not suit the couple who resettled in Florida. America's 'Ace of Aces' died of pneumonia in 1973 while on a visit to Switzerland.. Edward Rickenbacker endeared himself to millions of Americans in the course of a career that spanned two World Wars..

Frank Luke

Frank Luke (1897-1918) Medal of Honour

Frank Luke blazed a spectacular trail of aerial combats over the Western Front. In a period of only 17 days he became the second-highest scorer of the American aces and then in equally spectacular fashion his career ended. He was born on Phoenix Arizona in May 1897 and could well claim to have a German heritage as his mother had emigrated from Germany to America shortly after the American Civil War. Frank Luke was one of nine children and attended public schools at Phoenix and later college. He was active in college football, baseball and athletics and during summer vacation he worked in the copper mines at Ajo Arizona. By all accounts Frank had an aggressive yet fun-loving personality. He was also handy with his fists as well as being an excellent shot with a .45 revolver and a rifle.

In September 1917 following America's entry into the war Frank enlisted in the Signal Corps. Later he made a request to transfer to the Air Service where he commenced ground training at Austin Texas. On completion of flight training at San Diego, Luke was commissioned second lieutenant and sailed for France in March 1918. Advanced training continued at Issoudon, followed by an aerial gunnery course at Cazeau in southern France. To his disappointment Luke was assigned a position as ferry pilot. He acquired considerable experience in the role on a variety of aircraft but it was a posting to a combat unit that he repeatedly requested. This was finally granted and on 26 July he was assigned to the 27th Aero Squadron, a Spad unit based at Saints.

His arrival at the squadron was not well received, as its members considered him a conceited braggart and accepted the newcomer with reservation. He soon found disfavour with senior officers with his complete contempt of authority. Others saw in Luke an ideal combination of flying skill and offensive spirit. Never a good mixer, Luke struck up an unlikely bond with fellow pilot Joe Wehner. They became a formidable team on their balloon-busting, destroying two Drachens on 14 September. It was the start of a period in which 9 balloons fell to the teamwork of Luke and Wehner. The fact remains that Luke's victories were achieved by Wehner's selfless efforts to protect him from the defending fighters.

On 18 September whilst on patrol Luke spotted 2 balloons over Labeauville. He attacked and destroyed both, leaving Wehner as top cover. Meanwhile six Fokkers appeared on the scene and proceeded to harass the outnumbered Wehner. It was a desperate situation that was relieved by the timely return of Luke who despatched two of them. Thinking that Wehner had escaped, Luke headed for the German lines and near Verdun he sent a Halberstadt down in flames. Whatever exultation Luke felt for his string of successes was quashed when Wehner failed to return.

Luke was justifiably saddened by his death and was encouraged to take a spell of leave. He returned early and resumed combat flying. On September 26 he destroyed a balloon but his wingman Lt.Roberts was shot down and killed. Angered and depressed with their deaths Luke embarked on a two-day frenzy of unauthorised flights. Finally his C.O. grounded him until he could confirm to military discipline.

In defiance Luke took off in his Spad and headed for enemy territory. An order was issued that Luke be placed under arrest as soon as he landed. It was destined to be an order that was never carried out.

At sunset a Spad flew low over American H.Q. at Souilly and dropped a note that read; *Watch 3 Hun balloons on the Meuse. Luke.* As good as his word Luke shot down the first and while despatching the second he was severely wounded. He continued on to Milly and brought down the third, and before force-landing at Murvaux he machine-gunned German troops in the village. Barely conscious but still defiant, Luke ignored their demands to surrender He produced his service revolver and fired into their ranks until he fell dead from the return fire. Frank Luke was posted as 'missing in action' until his grave was located in a Murvaux cemetery; and later he was buried with full military honours at the American Military Cemetery at Romagne. His total number of victories was given as 21 and for actions at Murvaux he was awarded a posthumous Medal of Honour.

Bert Hinkler

Australia's Lone Eagle (1892-1933)

Born in 1892 at Bundaberg Central Queensland Herbert John Hinkler was the quiet achiever of Australian aviation and from an early age he displayed an interest in aeronautics. By the time he was twenty the diminutive Hinkler had built and flown a series of man-carrying gliders. In order to advance his aviation career he travelled in England and during World War 1 he served with the Royal Naval Air Service on the Italian Front.

Following the 1918 Armistice Hinkler settled in England and eventually he became test pilot for A.V. Roe, a leading aircraft manufacturer of the time. He was an interested observer of Ross and Keith Smith's 1919 England- Australia flight and was determined to emulate their feat but in a totally different manner. Instead of a large aircraft and elaborate preparation he opted for a miniature Baby Avro. It was a basic monoplane design powered by a modest 35hp engine, but it was the genesis of the future light aeroplane.

Despite official discouragement Hinkler set off from Croydon on 31 May 1920. He reached Turin non-stop in ten hours, which was a world record for a light aircraft. It was in Iraq that he fell victim to officialdom when authorities there refused him permission to fly over the Iraq desert. Hinkler's flight had officially failed but at least he had demonstrated the practicalities of light aircraft. Later in the year he shipped the Avro to Australia where he made a flight from Sydney to Bundaberg. It was a significant event, for he was able to fulfil a promise to his mother that one day he would fly home in his own aeroplane. He returned to England and continued his career as chief test pilot for Avro. A measure of his ability was his selection as reserve pilot for the 1925 British Schneider Trophy team. This was a prestigious international speed event for seaplanes, which was inaugurated in 1913 by Frenchman Jacques Schneider and following the war it was initially contested every year.

In 1925 it was held at Baltimore but Hinkler's assistance was not enough to win the trophy for Great Britain. Instead it was won by The United States and the winning pilot was the remarkable Jimmy Doolittle, who rose to the rank of general in World War 2. That same year the Avro Avian, a 2-seat biplane of metal construction and fabric covering went into production. Powered by an 80hp Cirrus engine it was capable of flying over 1,000 miles non-stop; as Hinkler demonstrated when he flew from London to Riga, Latvia; a distance of 1,200 miles.

On 7 February 1928 he embarked on his record-breaking flight to Australia. It was a great challenge as it represented the first realistic solo attempt. His first stop was Rome, one of many he made through extremes of bitter cold and merciless heat. He crossed the Mediterranean Sea to Tobruk and then on to Basra and the forbidding Arabian Desert. Hinkler reached Karachi on 14 February, setting up a new record from England to India.

Apart from a minor repair to an oil tank, the Avian was performing faultlessly. He completed the crossing of the Indian continent and on the next stages to Singapore and Java he experienced conditions of heavy rain and poor visibility. Possibly the most hazardous stage was the final one from Bima, an island east of Java.

From there he was faced with a 600 miles crossing of the lonely Timor Sea. There were no dramas and at 8pm on 22 February Hinkler landed at Darwin, fifteen and a half days after leaving England. His actual flying time was 134 hours or 5 days and 14 hours.

The quiet achiever had made the first solo flight from England and the first in a light aeroplane. Awards and acclaim were bestowed on the unassuming Hinkler. These ranged from a triumphal passage around Australia, an honorary rank of squadron leader in the RAAF to an award of the Air Force Cross from H.M King George V. Hinkler's record stood intact until May 1930 when Charles Kingsford Smith lowered the time to nine days and 22 hours. His machine was also an Avian, which he named 'Southern

Cross Junior'.

In 1930 Hinkler became a manufacturer in his own right when he designed and built a 2- seater wooden amphibian. It was named the Hinkler Ibis and was unique in having its two engines mounted in tandem above the single wing. Only one aircraft was built and after limited use it was stored for many years at his Southampton home. It was finally scrapped in 1959.

In 1931 Hinkler set out on the flight of his career, a journey from New York to London. His machine was a DH Puss Moth powered by a 130hp Gypsy Major engine. It did not possess the range to achieve a crossing of the North Atlantic but he achieved his goal with a novel approach to the problem. The Puss Moth also lacked certain equipment required for night flying over The United States, making it necessary to fly direct to Bermuda. This was an 18-hour flight and the first between the two cities. From there he flew south in easy stages from Venezuela to Port Natal on the Brazilian coast.

He was then faced with a 3,200 km ocean crossing to Bathurst on the west coast of Africa. Six hours into the flight Hinkler encountered fog and heavy rain, which persisted throughout the entire crossing. His aircraft was not equipped with blind-flying aids, forcing him to fly by compass and his own instincts; yet so accurate was his course that he made a landfall with an error of just one degree.

That stage was the first west/east crossing of the South Atlantic and in the process he made the longest non-stop flight in a light aircraft. Hinkler reached London on 7 December 1931, having flown 16,000 km. to achieve his goal, and in a quite unique manner.

But the skies that highlighted Hinkler's exceptional abilities were about to betray him. In January 1933 he set out in the Puss Moth in an attempt on the England-Australia record. At the time it was held by C.W.A. Scott in a Gypsy Moth in eight days and 20 minutes. On the same day after leaving

England Hinkler was reported over the Italian Alps in the vicinity of the Pratamagno Mountains. Following that sighting there was no further news until 1 May when itinerant forestry workers discovered wreckage of the Puss Moth. Nearby was Hinkler's body.

It was a sad finale to his career; one that he pursued with quiet determination and a preference to achieve those goals on his own. In recent years there were inferences of sabotage to the propeller of his aircraft at his last fuel stop. If such was the case it was the foulest of acts; the sad fact remained that the lone eagle had fallen. Bert Hinkler was later buried in the village cemetery in Pratamagno village. His achievements were not forgotten, in particular in Bundaberg where the original family home has been set up as a museum to honour his memory; also in a Brisbane museum where the trusty Avian is on permanent display.

Charles Lindbergh

Charles Lindbergh (1902-1974) The Spirit of St.Louis

Born on 4 February 1902 Charles Augustus Lindbergh Jr. was the son and only child of a lawyer and congressman. His early years were spent on the family farm at Minnesota; however things mechanical held more appeal than life on the farm. He studied mechanical engineering at the University of Wisconsin before embarking on an aviation career.

In 1923 he became one of those barnstorming daredevils who performed with their wing-walking and parachute jumps at local fairs and other venues. Having survived that period Lindbergh enlisted in the US Army in 1924 and trained as an Army Reserve pilot where he graduated at the top of his class. He later worked as a pilot in the embryo Air Mail Service, flying back and forth between St. Louis and Chicago. It proved to be highly dangerous tenure for those pioneer mail pilots, flying in war-surplus DH4 bombers in extremes of weather that were endemic to these areas. On three occasions Lindbergh was compelled to take to his parachute when weather conditions became too impossible to continue.

In 1919 history was made with the first non-stop aerial crossing of the North Atlantic. John Alcock and Arthur Whitten Brown were the airmen involved and their exploit ranged as one of great flights of the decade. Even greater interest was aroused in 1926 with the announcement of a substantial prize of $100,000 for the first non-stop flight New York to Paris. Businessman Raymond Orteig was the instigator of this prestigious event which attracted entries from both sides of the Atlantic.

Despite the daunting logistics of such a flight and lack of a suitable aircraft, Lindbergh was quietly determined to be a competitor. His precarious financial situation was fortuitously solved by the intervention of a panel of St. Louis businessmen who provided the necessary funds for the project. Unlike his rival competitors who favoured multi-engine types, Lindbergh opted for a single-engine solo approach. He approached the fledgling Ryan Company based in San Diego who undertook the construction of an aircraft suitable for Lindbergh's requirements, based on their successful M-2 design.

Lindbergh worked alongside the Ryan design team and their efforts were rewarded with a purposeful high-wing monoplane, powered by the iconic 400hp Wright Whirlwind. It was registered Ryan NYP (New York-Paris) and in recognition of his financial backers was named 'Spirit of St. Louis'. Meanwhile there was mounting excitement at New York's Roosevelt Field with final preparations for the event. By the time Lindbergh had flown cross-country from San Diego to New York, two earlier Atlantic attempts had ended in tragedy.

By 20 May 1927 the prevailing inclement weather pattern had improved to the extent to prompt Lindbergh to begin his odyssey. Conditions for the early morning take-off were far from favourable with low cloud and a waterlogged airfield. Despite the early hour a large crowd of spectators had gathered to witness the event as the 'Spirit' strove to gain momentum in the muddy conditions. Adding tension to the moment; its pilot was confronted by a line of power cables just beyond the limits of the take-off area, and then after a tentative hop or two the 'Spirit' was airborne, clearing the wires by the barest of margins.

Now facing the intrepid Lindbergh was a daunting challenge; a featureless stretch of ocean that was subject to extreme vagaries of weather. The young airman proved equal to the task, despite an overwhelming fatigue that threatened to engulf him. Finally, after 33.5 hours at the controls and with 3,600 miles behind him Lindbergh was in sight of his goal, the airfield at Le Bourget, Paris. An estimated 100,000 people were on hand to welcome the airman; to be part of a history-making event, possibly the flight of the century.

The taciturn Lindbergh had become an instant celebrity, feted by an enthusiastic public wherever he chose to appear. He devoted much of his time to the promotion of commercial aviation, at the same time authoring a succession of best-selling books. During a trip to Latin America he met Anne Morrow whom he married in 1929 and together they enjoyed the privacy that flying afforded them. In their quest for a life away from the spotlight the couple moved to a remote estate in Hopewell, New Jersey and in 1930 their first child, Charles was born.

At only 20 months old the boy was kidnapped from their home in 1932. The crime, with its ransom demands made headlines worldwide. Meanwhile the Lindberghs paid the $50,000 ransom but sadly their son's dead body was found in the nearby woods some days later. The police traced some of the ransom money to Bruno Hauptman, a carpenter with a criminal record and arrested him for the crime. The subsequent trial developed into media frenzy, with the police determined to convict Hauptman. Vital details which would have exonerated Hauptman were deliberately withheld during the trial until finally after four years of refused appeals the hapless Hauptman died in the electric chair.

In an effort to escape the constant media attention the couple moved to Europe, living at first in England and later France, where Lindbergh became involved in scientific research with a French surgeon. He also continued his work in aviation, serving on the Board of Directors of Pan American World Airways. Lindbergh was also invited by Reichsmarshal Hermann Goring to tour German aviation facilities; highly impressed by what he was shown he considered the possibility of settling in Germany.

However the couple returned to America; and convinced that in a future war Germany would be unbeatable Lindbergh became involved with the America First organization, which advocated that The United States stay neutral in the war in Europe. This attitude had the effect of eroding his public support, to the extent of suspecting him of Nazi sympathies. Following the attack on Pearl Harbour he became active in the war effort, working with Henry Ford on bomber production and acting as an advisor and test pilot for United Aircraft.

After the war he wrote several best-selling books, including 'The Spirit of St. Louis' which won the 1954 Pulitzer Prize for autobiography. In his later years Lindbergh moved to the Hawaiian island of Maui. He was to die of cancer at this remote retreat in August 1974, survived by his wife and five children. Despite past controversies Lindbergh is remembered for incredible feats of bravery and with helping to usher in the age of commercial aviation.

Charles Kingsford Smith

The Immortal Smithy (1897-1935)

First across the Pacific'. That was the burning ambition of ex-RFC pilots, Charles Kingsford Smith and Keith Anderson. In Western Australia in the early 1920s they operated an aerial taxi service and later a road haulage business. In 1927 the partners moved to Sydney where they hoped to find local air charter work. The venture was unsuccessful but a meeting occurred with a third party, which had a profound effect on their fortunes. Charles Ulm had no piloting skills; instead his talents lay in organizing, which he pursued in a ruthless, unwavering manner. The easy-going Anderson soon found himself relegated to the background by the thrusting and single-minded Ulm.

In an attempt to gain sponsorship for the Pacific flight, Smithy and Ulm embarked on a round-Australia flight using one of Smithy's elderly Bristol Tourers. The circumnavigation was completed in a record 10 days, which created favourable responses from various sources. Confident that the financial problems were covered, the trio boarded a steamer for The United States in July 1927 to prepare for the Pacific flight.

Shortly after their arrival they were able to purchase an aircraft suitable for their requirements from Polar explorer Hubert Wilkins. This was a Fokker FV11 B tri-motor, one of a pair that Wilkins had used on a recent Polar flight. The Fokker, which they named 'Southern Cross' came without motors and instruments. These were beyond their current funds, but with an additional one thousand pounds from the NSW Government and a further fifteen hundred pounds from a Melbourne businessman the project gained momentum.

'Southern Cross' was fitted with three Wright Whirlwind motors, instruments and long- range tanks, but the partners' money was fast running out. The situation became so desperate that Anderson returned to Australia, convinced that the project had no future. Smithy and Ulm were basically destitute and burdened with debts that amounted to $16,000. Reluctantly they made the decision to sell the Fokker and return to Sydney.

In a stroke of good fortune their financial problems were solved by the intervention of millionaire banker, Alan Hancock. The quiet businessman was also an experienced marine navigator with a great interest in aviation and was sufficiently impressed with the Australians to take over their sizeable debts and fund the Pacific flight. Two Americans were recruited to complete the crew; Harry Lyon, former ship's captain and a superb navigator, and James Warner, a radio operator with many years' service with the U.S. Navy and Merchant Marine.

Here was Smithy's moment of destiny; for years his obsession was the Pacific flight, and after incredible setbacks it was about to be realized, although it carried a bitter irony for Anderson, after being denied his opportunity to be part of the team. On the last day of May 1928 the grossly over-laden 'Southern Cross' lifted off from San Francisco's Oakland airport. The take-off was the most critical stage of the flight, so laden with fuel their speed was barely above the stall. It was a situation they faced for several hours until enough fuel was burned for them to gain precious altitude. Communication between navigator and pilot was achieved by way of notes pushed through a tube that ran from rear cabin to cockpit.

Engine noise was so intrusive that conversation between pilot and co-pilot was impossible. They too were compelled to resort to scribbled notes to each other The first stage of the journey was blessed with ideal weather conditions; clear and sunny and with the bonus of a tail wind. Twenty-four hours into the flight they were faced with a crisis when the batteries for the receiver and transmitter expired, making it absolutely vital that Lyon's navigation was faultless. Several times land was sighted, but to their disappointment the sightings proved to be cloud masses. In the midst of their problems Warner received a reassuring signal from a shore station giving them an indication of their position. At the same time Lyon managed to obtain an accurate sun shot that verified the signal.

For another hour 'Southern Cross' cruised above heavy cloud while the crew watched anxiously for a long-overdue sighting of land; becoming a race against the possibility of running out of fuel and a landfall. At last they were rewarded with the sight of the 14,000 feet volcano, Mauna Kea on Hawaii

Island. Lyon's navigation had been immaculate. Sixty minutes later they landed at Wheeler Field at Honolulu to complete the 2,300-mile crossing in 27 hours 30 minutes.

Following the euphoria of the first leg, 'Southern Cross' and its crew faced the most critical stage of the journey; 3,150 miles to Fiji and the first ever attempt by air. Smithy chose a beach at Barking Sands on the outer island of Kauai for the take-off. In retrospect it bordered on the suicidal with the all-up weight dangerously close to the aircraft's limits, the episode was a nerve-wracking experience for its crew and for the spectators. They watched in trepidation as 'Southern Cross' staggered into the air with its landing wheels skimming the waves as Smithy kept it airborne. It was some time before they clawed their way to an altitude of 500 feet and received assistance from a helpful tail wind. Three hours into the flight the troublesome radio problems returned, leaving them with only the capacity to transmit messages. Around mid-day they hit the inter-tropical convergence zone, an area that was host to storms of extreme ferocity.

Menacing clouds exploded with thunder and lightning, rising to altitudes of 35,000 feet; and quite impossible to fly over. 'Southern Cross' plunged into the boiling mass where it was flung about like a leaf in a storm.

All of Smithy's skills were needed to keep it airborne, in conditions that were akin to flying through a waterfall that drenched the pilots from head to foot. At one stage the aircraft was almost flipped upside down, so severe was the turbulence. Eventually they emerged into clear skies, and by Lyon's reckoning they had covered over 1,000 miles, almost one third of the distance. To their horror they saw ahead of them another threatening cloud mass, hundreds of miles across. Once more it became a desperate battle against the elements as Smithy endeavoured to climb above the worst of it. After three terrifying hours they emerged into a moonlit starry night. According to Lyon they were 1,100 miles from Fiji and with 500 gallons of fuel remaining.

By Smithy's calculations that was inadequate, which meant they might have to find an alternative island for a set-down. The remaining hours until daylight dragged by in an atmosphere of tension and when the time came to hand pump the remaining fuel they discovered there was ample to cover the distance. Lyon was able to obtain a shot of the sun during the morning, which showed they were north of their intended course. It was a tribute to his navigation after the constant detours during those violent storms and without a single radio bearing from ship or shore to assist him.

'Southern Cross' arrived over Suva in the early afternoon and Smithy was not impressed with the miniscule landing ground at Albert Park. He made a search for a suitable beach as an alternative but they were even less promising; it had to be Albert Park. An enormous crowd had gathered to witness their arrival and they were privileged to see Smithy perform a masterly landing in the confined space. Just when it seemed 'Southern Cross' was about to career into some large trees Smithy swung away in a spectacular ground loop. The Americans were crouched in the rear of the fuselage in an endeavour to hold down the tail of the aircraft. During the violent landing the unfortunate Warner was flung through the fabric and knocked unconscious. He was lucky not to be run over, and after being rendered first aid he quickly recovered.

For the final leg to Brisbane Smithy used Naselau Beach from where they took off on Friday 8 June. Being a far shorter flight; a maximum fuel load was deemed unnecessary. As they headed into a brilliant clear sky, crew members were quietly confident they were almost home. The difficult part of the journey was behind them, and besides they could hardly miss a target the size of Australia. Their euphoria was dealt a severe blow when soon after nightfall they flew into an electrical storm of terrifying proportions. All of Smithy's blind- flying skills were called on to keep 'Southern Cross' airborne. In an attempt to find a more favourable situation they climbed to 9,000 feet, where they faced bitterly cold conditions. Even at that altitude the Fokker was forced up and down in alarming surges. In the rudimentary cockpit the pilots were flung out of their seats as they clung desperately to the control wheels. The

Americans fared even worse in conditions that at times rendered them weightless.

For five hours of their ordeal it was impossible for Lyon to navigate and when they did emerge from the storm he estimated they were at least 100 miles off course, which was fortunate their goal was Australia and not a speck in the ocean like Fiji. Around 8am a shadow appeared on the horizon, which according to Lyon was the coast of northern New South Wales. Smithy identified it as Ballina, 110 miles south of Brisbane. An error of that magnitude on the Fiji leg did not bear consideration.

As they neared Brisbane, 'Southern Cross' was welcomed by a flight of light aircraft and at 10.15 a.m. Smithy made a textbook landing at Eagle Farm aerodrome. It would be difficult to put a precise number to the crowd that had waited since 3am. The figure varied from 15,000 to 40,000 but there was no doubting their adulation for the crew. Smithy became the centre of attraction as women attempted to smother him with kisses while men fought to shake his hand. Flight statistics showed that they had covered 7,220 in eight and a half days and in a flying time of 83 hours. Present day travellers might take a moment to compare their journey in the pressurised ease of a 747 Jumbo to the acute discomfort and appalling danger of that 1928 journey.

Richard Byrd

Richard Byrd (1888-1957)

The lure of adventure and exploration burns deeply in a select company of men and women. A prime example of this was Commander Richard Byrd who made a specialty of Polar exploration in both Arctic and Antarctic zones. He was the first to reach the North Pole and fly around it, yet in later years this achievement was clouded in controversy.

Richard Evelyn Byrd was born in 1888, the son of an influential Virginia family. His boyhood years were spent in outdoor activities, roaming the rugged hill country of North Virginia, sailing his skiff along the Chesapeake, swimming, riding and long-distance running. Byrd's career aspirations were focussed on joining the U.S. Navy and in this he was successful, following his entry to the U.S. Naval Academy in 1912 where he proved to be a fine student and good athlete.

However a foot injury caused his retirement after three years active service, nevertheless he was able to join the U.S. Navy's fledgling Air Arm where he developed into a proficient pilot. Byrd also took a keen interest in navigation and was responsible for the creation of the 'bubble horizon' for the sextant, the drift meter and the sun compass. As a consequence he was involved in a technical and administrative capacity in the preparations for the first transatlantic crossing by a team of U.S. Navy flying boats in 1919.

In that post-war period Byrd became part of the political struggle between the U.S. Army and U.S. Navy over support for aviation. The Army Air Corps, backed by adulation for its wartime service in France was conducting a campaign to gain Federal budget support. The Corps' principal opponents were the 'Battleship Admirals', using the influence of Admiral Moffett, himself a staunch proponent of aviation, including the employment of airships in naval strategy.

Byrd appreciated the need for public support in the Navy's bid for budget assistance and accordingly he sought permission for a flight over the North Pole as an obvious means of gaining publicity. This was not forthcoming and he had to settle for the more mundane task of providing light air support for naval expeditions in northern waters. In 1924 he had command of a small naval detachment accompanying an arctic expedition to northern Greenland.

Frustrated over the controversy surrounding the claimed success of early American flights over Greenland and Ellesmore Island, Byrd applied for an extended leave of absence from the Navy. His intention was to arrange a private expedition to fly over the North Pole, financed by automobile magnate Henry Ford. Byrd was determined to prevent the renowned Norwegian polar explorer Roald Amundsen from becoming the first to reach both Poles. Amundsen was already planning to overfly the North Pole in the dirigible *Norge*. The airship was built in Italy and was already on its way by sea to Trondheim, from where it would fly on to King's Bay, Spitzbergen and from there on to the Pole.

The situation was developing into a desperate race to be the first to make that historic aerial crossing. Byrd's preparations were well in hand with his Fokker tri-motor *'Josephine Ford'* already shipped to Spitzbergen. As the aircraft was being unloaded on 9 May 1926 the *Norge* arrived in King's Bay; if Byrd's attempt was to be successful there was clearly no time to lose.

Only a few hours later the *Josephine Ford,* with Floyd Bennett, an enlisted pilot/mechanic at the controls and Byrd navigating, took off for their historic 15½ hour flight which included four minutes flying around the Pole. On their return to The United States both airmen were greeted with great adulation and in Byrd's case a promotion to commander.

Their next project was to compete for the $100,000 prize for the first non-stop flight from New York to Paris. Backed by the Wanamaker clothing concern Byrd's attempt would be in another Fokker, *The America*. However during flight testing at Roosevelt Field the aircraft overturned on take-off. Bennett was badly injured while Byrd escaped relatively unhurt and before a new attempt could be made Charles Lindbergh had already claimed the substantial prize.

On 29 June 1927 Byrd took off for Paris with the first official mail service, piloted this time by Bert Acostia, Their actual crossing passed without incident, however after arriving over France heavy cloud caused them to abort their landing at Paris. Instead they turned back to the Normandy coastline to attempt a landing at Ver-sur-Mer and in the ensuing crash landing their Fokker suffered some damage, fortunately without serious consequences for its crew.

In 1928 Byrd directed his exploration to the Antarctic zones. This expedition was particularly well prepared; including two ships and three aircraft. One of these was a Ford tri-motor, named Floyd Bennett after the recently deceased pilot of Byrd's earlier expeditions. A base camp named 'Little America' was set up on the Ross Shelf, from where scientific expeditions were undertaken during that summer. The highlight of the expedition was Byrd's successful flight over the South Pole, using the Ford tri-motor piloted by Bernt Balchin and co-pilot Harold June. Some difficulty was experienced in gaining adequate altitude over the Polar Plateau; in this they were successful and following an 18 hour flight the 'Floyd Bennett' made a triumphant return to 'Little America'.

During Byrd's second Antarctic expedition in 1934 he spent five winter months alone operating a meteorological station named 'Advance Base'. He narrowly escaped with his life after suffering carbon monoxide from a poorly ventilated stove. Two rescue attempts from their base camp were thwarted by bad weather and mechanical problems until finally a third mission arrived at Advance Base where they found Byrd in poor physical health. In October Byrd was flown back to base while the rest of the team returned by tractor.

Byrd's third and final Antarctic expedition was the first to receive official U.S. Government backing. The project included extensive studies of geology, biology and exploration. Within a few months in March 1940 Byrd was recalled to active duty in the Office of the Chief of Naval Operations, while back in the Antarctic the expedition continued without him.

From 1942 to 1945 he headed important missions to the Pacific, including surveys of remote islands for future airfields. After a lifetime of exploration and adventure the intrepid Richard Byrd died peacefully in his sleep at his Boston home on 11 March 1957.

Jimmy Doolittle

Jimmy Doolittle (1896-1993) Tokyo raider

In a lifetime encompassing aviation James Harold Doolittle emerges as a truly larger than life character. He was born in 1896 in Alameda near San Francisco; however his formative years were spent at Nome, Alaska where his father worked as a carpenter. It was a harsh, lawless society and by the time the family returned to Los Angeles the young Doolittle was well able to look after himself.

His first contact with aviation was at a local air show in 1910, which inspired Doolittle to experiment with home-built gliders. In 1917 while he was studying for an engineering degree America entered the war, prompting Doolittle to enlist in the Air Service. He had no opportunity to see action overseas; however in a post war situation he became the first pilot to cross America in less than a day. This was achieved in a DH4, crossing the continent from Pablo Beach Florida to San Diego, a distance of 3,581 kms in a flying time of 21 hours.

He then returned to his studies, gaining a degree in Aeronautical Science, and from there he took a course in seaplane flying at Anacostia Naval Air Station. Doolittle consequently became a member of the United States team to contest the 1925 Schneider Trophy races held that year at Maryland. It proved to be a resounding triumph for Doolittle in the Curtiss R-C3 floatplane to beat his British and Italian rivals, recording a world seaplane record of 374kph in the process.

Doolittle's technical expertise and flying skills made him an ideal demonstration pilot for the Curtiss-Wright company. In 1926, after receiving the Mackay Trophy for services to aviation he made a tour of South America, demonstrating the Curtiss-Wright P-1 biplane. A similar tour was carried out in 1928.

Following a secondment to the Full Flight Laboratory Doolittle pioneered the development of blind-flying aids, then in its infancy. In 1929 with these instruments fitted to a Consolidated NY-2, plus a crude type of radio beam device Doolittle made the first-ever 'under the hood' flight from Mitchell

Field New York. In case of trouble a check-pilot accompanied Doolittle on this ground-breaking epic.

In 1930 he resigned his commission to work for the Shell Oil Company in St. Louis, but was soon released for another Curtiss-Wright sales tour, this time in Europe. Air racing was then at its height of popularity, which saw Doolittle adding the Bendix and Thomson Trophies to his collection. The latter was achieved in the wildly-unpredictable Granville GB Special at a world record speed of 476kph and on that note Jimmy Doolittle retired from air racing. He was one of the few pilots to fly the notorious GB and live to tell the tale.

Following two trips to Europe in 1937 and 1939 on behalf of Shell Oil Doolittle was alerted to the inevitability of a Nazi-inspired European war. In July 1940 he was recalled to military service with the rank of major. His initial brief was to supervise the conversion of Detroit car makers to aircraft manufacture. His goal however was an operational flying posting and his appeals for transfer at first went unheeded.

This all changed dramatically in January 1942 when Doolittle was summoned to the office of Major General 'Hap' Arnold, chief of the USAAF. It was the creation of 'Special Aviation Project No.1', a highly-secret mission to strike at Japan with carrier-based aircraft. The North American B25 Mitchell had recently entered service and was selected by Doolittle as the most suitable for the daring strike.

The mission was organised on a volunteer basis, using crews from 17th Bombardment Group and 89th Reconnaissance Squadron. Training began at Eglin Field, Florida practising short take-offs from an airfield laid out with lines and flags to represent the flight deck of an aircraft carrier. Certain equipment, such as the ventral gun turret and top-secret Norden bomb sight were removed and engine carburettors adjusted for minimal fuel consumption.

Training completed; 16 aircraft and crews embarked on the carrier Hornet, which sailed from San Francisco Bay on 12 April 1942. General Arnold held misgivings about Doolittle leading the raid, due to the extreme risks involved, however by dint of persuasion he ensured a place in the lead aircraft. Once at sea the crews were informed of their objectives; military installations at Tokyo, Yokahama, Osaka and Nagoya.

Scheduled take-off was mid-morning 18 April when the carrier and its escorts would be 400 miles off the Japanese coast. A crisis arose at 6.00am when a Japanese picket boat sighted the task force. The vessel was quickly sunk by the escorting destroyers but it was now imperative that the bombers be despatched, although they were still about 650 miles from the mainland.

At 8.00am Doolittle in the lead bomber began his take-off, engines bellowing as he moved along the Hornet's pitching deck, and then with just metres to spare the heavily laden B25 'un-stuck' and lifted into the leaden atmosphere. It was a heartening experience for the remaining 15 crews who were all flown off without major incident.

With one exception all crews managed to locate and bomb their assigned targets. One aircraft was engaged by Japanese fighters and had to jettison its load, yet by and large the bombing results were reasonably successful. Their next challenge was to locate the emergency Chinese landing field at Chuchow, but with no radio assistance and short of fuel there was no alternative than to bail out or ditch off the coast. Doolittle, like most crews chose to bail out over China and on the following morning he and his crew were rescued by Chinese soldiers and eventually re-united with five other crews.

Two crews fell into the hands of Japanese sympathisers and Doolittle's attempts to free them through a local governor failed. Another B25 landed in the USSR where its crew was interned by the Russians. Japanese retribution for the raid was swift and brutal, with an estimated 120,000 Chinese being murdered for any complicity in aiding the American airmen.

Analysed coldly the mission was hardly successful, with all 16 aircraft being lost and seven aircrew killed. Post-war evidence suggests that the Japanese believed that the raid was launched from Midway, thus precipitating their ill-considered attack on June 3 1942. In the ensuing action they lost an estimated 330 aircraft, five carriers and a heavy cruiser.

Midway was considered to be a turning point in the Pacific war, so the sacrifices of the Doolittle raiders were not entirely in vain.

Doolittle returned to the US as a newly-promoted Brigadier General to be awarded the Congressional Medal of Honour for his part in leading the 'Tokyo Raiders'. His career continued, first with command of the Twelfth Air Force in North Africa in 1942, the Fifteenth Air Force in Italy in 1943 and the Eighth Air Force in England. Following VE Day he took the Eighth to the Pacific, where conversion to the B29 Superfortress began, however few missions were undertaken before the Japanese surrender.

Post-war Doolittle re-joined Shell Oil in 1946 as Vice-President, while still maintaining his army reserve status. He resigned from the service in 1956. Prior to his death at age 96 in Pebble Beach California on 17 September 1993 he presided at annual reunions of the surviving 'Tokyo Raiders. At the time of writing just a handful of these remarkable airmen still survive; from an operation for which Jimmy Doolittle will ever be remembered.

Amelia Earhart

Amelia Earhart (1897-1937)

In the course of an exciting and at times turbulent career Amelia Earhart advanced the cause of female aviators more than any other woman. At the apex of a ground-breaking career her star was extinguished and when the end came it was a situation of Deja vu, as Amelia had once said,' When I go out I should like to go quickly and in my own plane.' The chances are that she did.

Amelia Mary Earhart was born 24 July 1897 in Kansas into a family situation that offered privilege and wealth through her maternal grandparents. However her father Edwin's faltering law practice saw a change in their circumstances when the family moved to Des Moines Iowa and for a time their living standards improved. This was merely a prelude to a disintegration of the family unit with Edwin's alcohol dependence, until finally in 1914 Amy Earhart and the girls went to live in Chicago.

In 1917 Amelia made the decision to train as a nurse's aide at a Toronto military hospital, which proved to be a sobering episode to witness at first hand the result of four years of war; nevertheless the experience inspired Amelia to enrol at Columbia University as a pre-med student. This was in late 1919 and although doing well in her studies, in 1920 she abandoned the course and joined her parents in California following a recent family reunion. It proved to be a propitious move for Amelia in respect of an aviation career for shortly after her arrival she went aloft for the first time in a ten-minute flight over Los Angeles. Exhilarated by the new experience Amelia was convinced that she was destined to be a pilot.

Soon she began lessons with a pioneer aviatrix Anita Snook at Kinner Field near Long Beach although her tutor had reservations about Amelia's natural piloting skills, a feeling later shared by many of Amelia's contemporaries. Over the following years Amelia was appointed to various aviation associations which gave her the opportunity to advance women's involvement in the field of aeronautics. These were still difficult years for Amelia in financial terms but in 1928 her situation improved markedly with the offer of becoming the first woman to fly the Atlantic.

This project was the brainchild of wealthy American socialite Amy Guest who had purchased a Fokker FVII tri-motor in which to make the attempt. Mrs. Guest's family objected vigorously to her involvement but later relented, provided the 'right sort' of woman could be found to make the flight. This was a challenging profile for any woman and on 27 April 1928 Amelia received a phone call from a Captain H.H. Railly who had been asked by New York publisher George Palmer Putnam to find such a person. Railly had been so impressed by Amelia's strong resemblance to Charles Lindbergh that the name 'Lady Lindy' was coined. A week later Putnam and Amelia met in New York and the publisher had no hesitation in making her a crew member although her role would merely be as a passenger.

An experienced duo, Wilmer Stultz and Louis Gordon would pilot the Fokker tri-motor, named 'Friendship' and now fitted with floats for the occasion. After several days of waiting for clearing weather conditions 'Friendship' departed Halifax, Nova Scotia on 18 June. Dense fog was encountered for most of the journey until a landfall was made in South Wales and not Ireland as planned, in a time of 20 hours 40 minutes.

In true press fashion Amelia was besieged by reporters while for the most part Stultz and Gordon were ignored, a factor that deeply distressed Amelia. From then on she was in great demand on the lecture circuit while behind the scenes George Putnam kept Amelia's name in the forefront of everyone's focus. Their close relationship had not gone unnoticed, particularly by Mrs. Dorothy Putnam who filed for divorce. This was granted and in February 1931 after a series of endurance and speed flights by Amelia and numerous proposals from George Putnam the couple eventually married.

It could be assumed that this was a marriage of convenience, giving Amelia the opportunity to purchase a state-of-the-art aircraft, an iconic Lockheed Vega. Determined to be the first woman to fly solo across the Atlantic she made plans accordingly, departing Harbour Grace Newfoundland on 21 May 1932. Her journey lasted almost 15 hours when she landed slightly off-course in a meadow in Londonderry Northern Ireland. Once again Amelia became the focus of intense media attention, culminating in a ticker-tape parade in New York. Her Atlantic triumph was a precursor to a series of record-breaking flights, notably as the first woman to make a transcontinental U.S. flight and as the first person to achieve a solo flight from Hawaii to California.

In 1937 Amelia began planning an around-the-world flight, following closely to the equator route. The faithful Vega had been replaced by a custom-built Lockheed 10E Electra fitted with extra-large fuel tanks. This was a far more appropriate machine for such a venture and it would not be a solo effort, as Amelia had enlisted Paul Mantz as co-pilot and Harry Manning as navigator. A west-bound route was chosen, departing Oakland 17 March and a first let-down at Hawaii. However on take-off from Luke Field Hawaii she over-compensated for a dropped left wing and the plane swung out of control. At that point the undercarriage collapsed and the fuel-laden Electra slid along the runway on its belly. Miraculously there was no fire but the aircraft had suffered a considerable amount of damage, leaving Amelia no alternative than to ship the plane back to the Lockheed where it became the subject of a major re-build.

The venture had proved to be a disappointing and expensive outcome for Amelia but later in 1937 she embarked on a second round-the-world flight, this time in an easterly direction with Captain Fred Noonan as navigator and with no co-pilot. They departed Oakland 21 May, crossing the southern United States to Miami prior to a South Atlantic crossing to West Africa via Puerto Rico and Brazil. Karachi was reached on 16 June, making it the first non-stop flight from the Red Sea to India. Monsoonal conditions were experienced during the next stage from Calcutta through Rangoon, Singapore and Bandung. At Port Darwin Noonan carried out repairs to the direction finder and from Darwin they proceeded to Lae on the north-east coast of New Guinea, having covered 22,000 miles during their odyssey. Their next goal was Howland Island, an ocean crossing of 2,550 miles and little more than a sandbar in a featureless expanse of water. A runway had only recently been constructed there and this stage of the flight would never have been contemplated without the presence of the U.S Coastguard cutter *Itaska* standing offshore to give radio assistance.

They departed Lae at 10 am local time and with an anticipated headwind the flight was expected to take between 17 and 20 hours, they had fuel for 24. Throughout the night the *Itaska* endeavoured to make contact with the Electra but static interference blotted out their signals. At 6.15am Howland time Amelia's voice at last penetrated the static. She informed *Itaska* that they were about a hundred miles out and requested a bearing on the Electra. There was no further contact until at 7.42 Amelia's voice came through with the graphic statement: *We must be right on top of you but we can't see you.*

But this brief interlude of improved reception ended; suffocated by the relentless static, until 45 minutes later Amelia's voice broke through. She stated their line of position and that only 30 minutes of fuel remained, which proved to be her final transmission. An immediate air and sea search began but the aviators were in an unknown position and despite a most comprehensive search by U.S Navy ships and aircraft no trace was ever found. Amelia had unwittingly achieved more fame in death than in life and her disappearance became the subject of a host of theories. Most of these were fanciful; that she was on a spy mission ordered by President Roosevelt or she was captured by the Japanese and forced to broadcast to American GIs as 'Tokyo Rose' during World War II. In a post war world researchers have maintained their investigations, with vague assertions as to her fate. A far better outcome would be to let her rest in peace beneath the pristine waters bordering that elusive Howland Island.

Jean Batten

Jean Batten (1909-1982)

In the decades following the Great War aviation was fast becoming a viable transport alternative. New airline routes were being established thanks to the dedicated efforts of the earlier pioneering trail blazers, typified by Charles Lindbergh, Charles Kingsford Smith and Alan Cobham, to mention just a few. At the same time female aviators were leaving an indelible impression on the record books. Three of those deserve special mention, Amelia Earhart, Amy Johnson and New Zealand's greatest aviator Jean Batten. Earhart and Johnson were destined to make the ultimate sacrifice in their dedication to a chosen career. Batten on the other hand was to die in virtual obscurity and in toxic circumstances.

Jean Gardner Batten was born in Rotarua in September 1909, which could be considered a prophetic omen for her future aviation career as this was the year that Frenchman Louis Bleriot made his historic English Channel crossing. Jean's father Frederick was a successful dentist and on the outbreak of World War I he enlisted in the armed forces and later served on the Western Front. This disruption to the family situation created financial hardship to Mrs. Ellen Batten and to Jean and her siblings. They were obliged to lead an itinerant existence in a succession of boarding houses until Fred Batten's return from active service. By that stage Ellen Batten was reluctant to relinquish her position as head of the house and around 1920 they separated permanently.

Jean opted to live with her mother with whom she developed an intense personal relationship that endured throughout their lives. During her time at a ladies' college Jean excelled in various subjects and had matured into a healthy and attractive young woman. Already she was regarded as a loner, an intelligent yet remote person whom few could warm to. With a view to perform professionally she began to study piano and ballet. However in May 1927 her ambitions changed dramatically, inspired by Charles Lindbergh's New York to Paris solo flight she yearned to be a pilot herself. It was a momentous decision, one that was encouraged wholeheartedly by Ellen Batten

In 1929 Ellen and Jean sailed to Australia where Jean had the opportunity to fly with Charles Kingsford Smith in his iconic Fokker tri-motor 'Southern Cross'. The experience reinforced her determination to become a pilot. This was achieved in 1930 by the expedient of selling her piano to pay the sea-fare for mother and daughter to sail to England. Jean learnt to fly at the London Aeroplane Club, gaining her 'A' license in December. Almost immediately she began planning for an attempt on the England- Australia women's record, then held by the English pilot Amy Johnson in a time of 19½ days.

Her current finances were insufficient to achieve such a goal and as a result Jean and her mother returned to New Zealand to raise funds for the project. However this proved unsuccessful, prompting Jean to return alone to London where she stayed with her brother John who was gaining a reputation as a movie actor. It was a bitter-sweet reunion, for shortly they quarrelled and Jean moved out, never to speak to one another again.

Hoping to attract a corporate sponsor in difficult financial times Jean decided to train for a commercial pilot's license. To fund the £500 required to complete the 100 hour course Jean began a relationship with Fred Truman, a New Zealander serving with the RAF. Truman was keen to marry the tyro aviatrix and accordingly loaned her the money. On gaining her 'B' license in December 1932 she calmly walked out of his life, making no attempt to repay him. She then transferred her affections to Victor Doree who was also infatuated with her. He was the son of a prosperous linen merchant and once again Batten managed to borrow a substantial amount of money. This time it was £400 which Doree's mother provided and with these funds she was able to purchase a Gypsy Moth to launch her on a record-breaking career.

Her inaugural record flight began on April 1933 in an attempt to better Amy Johnson's time to Australia. It was plagued with incidents of frightening proportions; forced down in a sandstorm over Iraq she landed in the desert and was forced to remain there until daylight. Resuming her flight she experienced engine trouble near Karachi and in the ensuing forced landing she was fortunate to survive, although the machine was wrecked. Back in London Jean approached Doree to buy another aircraft and not surprisingly he refused, thus ending a shaky relationship.

Undaunted, Jean then made an approach to Charles Wakefield, head of Castrol Oils who had been impressed by her determination and glamour. Boosted by his sponsorship she purchased a second-hand Gypsy Moth for £240. Batten was now engaged to Edward Walter, a London stockbroker and in April 1934 she set out on a second England-Australia attempt. This too ended in disaster; out of fuel on the outskirts of Rome she encountered a maze of radio masts. With great skill she force-landed and in the process suffered painful facial injuries.

After repairs to the Moth she flew back to England to prepare for a third attempt. With a set of lower wings borrowed from Walter's personal aircraft Jean made final preparations and on 8 May she set out again. This time her flight was crowned with success, arriving at Darwin in 14 days 22½ hours. In the process Amy Johnson's time was slashed by over 4 days and overnight Jean Batten had become a world celebrity.

Lionised by the Australian and New Zealand public she revelled in this new-found adulation. She embarked on lecture tours, impressing with her poise and oratory. A pleasing bonus to this was a welcome period of prosperity, to compensate for those earlier years of austerity. Her romantic situation had altered dramatically following her affair with Beverly Shepherd, an Australian airline pilot. This was not well received by Walter who angrily demanded compensation for the set of wings he had lent her. Meanwhile Jean flew the Moth back to England, thus becoming the first woman pilot to fly from Australia to England.

She was now hailed as an international celebrity, with her achievements being widely recognised by the press. Not content to rest on her laurels Jean purchased a new cabin monoplane, a Percival Gull 6 for £2000, a sizeable sum in 1935. In November she flew from England to South America via West Africa, in what was recognised as a brilliant feat of navigation.

Her passage from West Africa to Brazil made her the first woman to pilot herself across the South Atlantic. In Brazil, Argentina and Uruguay Jean was the object of adulation from presidents and public alike. More honours followed, being awarded the Royal Aero Club's Britannia Trophy and jointly awarded with Amelia Earhart the Harmon International Trophy for the most outstanding flight for a woman in 1935. Now based in England Jean and Ellen rented a cottage in Hertfordshire and for a period they virtually disappeared from society.

In October 1936 Jean emerged from her seclusion to make the longest of all her odysseys, the first-ever flight from England to New Zealand. On 16 October she landed at Auckland's Mangere airport to be greeted by an estimated 6000 people. The epic journey had taken 11 days 45 minutes, however the effort had drained her mentally and physically and in February 1937 she returned to Sydney to be reunited with Shepherd. It proved to be a tragic outcome for on the very day she arrived; Shepherd was killed in an airliner crash in the MacPherson Ranges of northern New South Wales.

Deeply grieved with his death it was not until October that Jean took to the air again. In a remarkable time of 5 days 18 hours she made the return journey to England, becoming the first person to simultaneously hold the England-Australia record in both directions. It would appear to have been her finest hour, and then quite swiftly she faded from the public gaze.

From then until the outbreak of war Jean and Ellen continued their itinerant life style, moving from one secret address to another around the United Kingdom. Jean was in Sweden when war was declared in September 1939 and received special permission to fly the Gull back to England over foreign air space. Back in the UK Jean offered her services to the ATA as a ferry pilot, with the proviso that she could still fly the Gull. When this was refused she declined to join the Auxiliary and for a time she served as an ambulance driver. Once again she was linked romantically to an airman, in this instance an RAF bomber pilot, but unfortunately he was killed in a mission over Europe.

In a post-war situation Jean and Ellen settled in Jamaica for seven years. From there they returned to England; to then spend a nomadic six years touring Europe until 1960 when they bought a villa at a fishing village near Malaga. It was in July 1966 while the pair was holidaying at Tenerife that Ellen died, aged 89. Inconsolable with grief Jean settled in Tenerife for the next 16 years where she led her typical isolated life style. There was a dramatic change of circumstances when in 1969 she returned to England and immersed herself amid much publicity in the Aviation world. This sojourn lasted for 12 months or so and then she flew to New Zealand for a brief visit before returning to Tenerife.

For the ensuing ten years Jean continued her restless odyssey around England and Europe before finally settling in Majorca. She died in tragic circumstances in November 1982 aged 73. This followed being bitten by a dog on one of her daily walks. Jean stubbornly refused to allow staff at her small hotel to call a doctor and as a result she died quite needlessly of blood poisoning. Unaware of her identity the local authorities buried her in a mass pauper's grave in Palma cemetery.

Eventually the New Zealand government overcame its inertia and re-interred her body in her home country. It was a sad conclusion to a brilliant if brief career, of the greatest female pilot of the Golden Age of aviation.

Adolph Malan

Adolph Malan (1910-1964)

During the period leading up to and including the Battle of Britain Adolph Malan gained fame as an outstanding fighter pilot and later a group commander with RAF Fighter Command. Born in Wellington, South Africa in 1910 his teenage career was focused on the merchant navy; hence the RAF nickname 'Sailor'. By 1930 he had gained his second mate's certificate and completed many transatlantic crossings

In 1935 he made a decision that would change his fortunes completely when he was accepted for pilot training with the RAF; a time of great expansion in the service. Commissioned in 1936 Malan was posted to No.74 'Tiger' squadron, a unit with a proud heritage begun in World War I. Its outstanding member was Major Edward Mannock who was destined not to survive the conflict but nevertheless was the highest-scoring of the British aces with 74 victories. In 1936 the squadron was operating with the biplane Gloster Gauntlet, which at the time was the RAF's fastest interceptor. But with war imminent their classic silver biplanes were soon to be replaced by the iconic Spitfire.

During the early months of World War 2 the UK-based Spitfire units saw little action until the onset of 'Operation Dynamo' in late May 1940, where over 300,000 British and French troops were retrieved from the beaches of Dunkirk. No.74 was one of the first Fighter Command units to be thrust into action to cover the evacuation. It was a period of intense aerial activity, during which Malan was credited with the destruction of two Ju88s, one He III, one Do17, and significantly a Me109, a type that would figure largely in his ultimate victory tally.

In June 1940, following a brief rest at Leconfield in Yorkshire No74 squadron returned to Hornchurch, with Malan as its commanding officer, he having been advised of his award of the DFC. In that period Malan undertook night-fighting patrols and became the first single-seater pilot to shoot down a bomber at night, with the destruction of a Heinkel III on the night of 18/19 June. Several nights later he was credited with two more Heinkels; with this remarkable achievement being recognised with the award of a Bar to his DFC.

July 1940 saw an increase in Luftwaffe activity with attacks on coastal shipping and other targets. After successes against enemy bombers earlier in the month, the RAF's main opponents became German fighters, and as a result some hard and costly combats were fought. Throughout this desperate period between July and August Malan claimed the destruction of five Me109s and two Dornier 17s. On August 12 the squadron was withdrawn to the Midlands to rest and reform and it was while there that Malan formulated his 'Ten rules of Air Fighting' doctrine. Its impact was well noted and shortly these precepts were displayed prominently at most fighter airfields and training bases.

In October the squadron returned to operations, this time to Biggin Hill, a base that it shared with the equally-famous 92 squadron. This was also a Spitfire unit and in contrast to Malan's strict disciplinarian tactics No.92 was somewhat easy-going and manned by characters with independent attitudes.

With the onset of autumn the large-scale daylight bombing had eased to a certain extent, nevertheless the intruding Me109s were still acting aggressively. Malan's personal tally was nearing twenty and later in December he was awarded the DSO and promoted wing commander (flying) at Biggin Hill.

The spring of 1941 saw a change of Fighter Command tactics with offensive sweeps over the French and Belgian coasts. These sorties would involve losses for both the RAF and Luftwaffe and with the launch of 'Operation Barbarossa', the German invasion of Russia in June 1941 there was a diminishing level of aerial activity for RAF Fighter Command. In that hectic period Malan was heavily involved as a wing leader, when he

claimed the destruction of 14 Me109s, taking his tally to 32 confirmed, which at the time made him the top-scorer in Fighter Command.

In October 1941 Malan accompanied Group Captain Harry Broadhurst and other luminaries of Fighter Command on a lecture tour of The United States, where Malan had the opportunity to fly the P38 Lightning and P39 Airacobra.

On his return to England in November Malan was promoted to group captain and became gunnery instructor at Central Gunnery School at Sutton

Bridge. He returned to Biggin Hill in January 1943 as station commander and presided over the celebrations when the famous airfield claimed its 1000th victory. Twelve months later he became commander of a wing of the new Second Tactical Air Force and was able to make some flights over the Normandy beachhead following the invasion in June 1944. Try as he might, a flying whenever the opportunity presented itself he was not to score again.

Malan's next posting, in July 1944 was to command the Advanced Gunnery School at Catfoss, where several other famous aces served under him as instructors. One of these was Pierre Clostermann, the top-scoring French pilot serving in the RAF; another was the celebrated American Richard Bong who claimed 38 victories in the Pacific flying the P38 Lightning. Malan then underwent a six-month course at the RAF Staff College during 1945 and in 1946 he resigned his commission and returned to South Africa.

For a period he became secretary to diamond millionaire Harry Oppenheimer and in 1950 he embarked on a career as a sheep farmer in Kimberley. Malan became a strong opponent of the new apartheid policy and a proponent of the Coloured Voting Bill; later becoming national president of this society. Sadly the great warrior's health was in decline with sclerosis and in 1964 he succumbed to this cruel disease after many months in hospital.

Malan's Ten Rules of Air Fighting

1. Wait until you see the whites of his eyes. Fire short bursts of 1 to 2 seconds and only when your sights are definitely "ON."
2. Whilst shooting, think of nothing else; brace the whole body; have both hands on the stick; concentrate on your ring sight.
3. Always keep a sharp lookout. "Keep your finger out."
4. Height gives YOU the initiative.
5. Always turn and face the attack.
6. Make your decisions promptly. It is better to act quickly even though your tactics are not the best.
7. Never fly straight and level more than 30 seconds in the combat area.
8. When diving to attack always leave a proportion of your formation above to act as top guard.
9. INITIATIVE, AGGRESSION, AIR DISCIPLINE and TEAMWORK are words that MEAN something in air fighting.
10. Go in quickly. Punch hard. Get out!

Pat Pattle

Pat Pattle (1914-1941)

Squadron Leader M.T. St. John Pattle was an outstanding fighter pilot and with some justification could well be regarded as the top-scoring Allied ace. Unfortunately all his combats took place in the Middle East and the Balkans, areas isolated from admiring press releases and resultant fame. More drastically the records of No.33 squadron were destroyed during the 1941 Greek campaign and evacuation, thus depriving authorities of essential details of Pattle's magnificent fighting career.

The future ace was born in Butterworth, South Africa in July 1914 and spent his early years on his father's farm at Keetmanshoop. It was there that the young Pattle developed his marksmanship skills while on hunting expeditions in that area. On leaving school Pattle applied to join the South African Air Force but to his disappointment he was rejected and instead he found employment as a clerk at a gold mine. He also joined the Special Service Battalion, a paramilitary unit through which he hoped would provide a means of entry into the SAAF.

However his aviation career came about through a newspaper advertisement urging young men to join the RAF. Seizing the opportunity Pattle sailed to England in April 1936 and was duly accepted for pilot training. On completion he was posted to No.80 squadron, originally a World War 1 unit and recently re-formed. Their aircraft was the biplane Gloster Gladiator, although an interim type pending the introduction of the Hurricane and Spitfire it represented the RAF's front-line equipment at the time. The squadron's UK tenure was relatively brief, for shortly afterwards it was posted to the Middle East, arriving in Egypt in April 1938.

Although the area was not officially at war there still existed a strong undercurrent of anti-British sentiment. More pertinent however was the build-up of Italian land and air forces in Libya, at that time an Italian colony. Their fascist dictator Benito Mussolini envisaged a re-birth of Italian glory in the Mediterranean theatre in the mould of Julius Caesar and his contemporaries. Already Mussolini had earned world-wide condemnation with his invasion of a helpless Abyssinia where tribesmen faced a modern army equipped with tanks, artillery and aircraft. Altogether it was a dastardly action, made even more so by the use of poison gas by the conquering Italians,

With no involvement in the Abyssinian debacle, British forces in the Middle East acted as policemen in the control of dissident Arab tribesmen in Palestine. It was an opportunity for Pattle and his fellow pilots to gain some early operational experience before the inevitable entry of Italy into the war. Hostilities began in July 1940 when Italian land and air forces launched their campaign against General Wavell's Desert Army and although numerically inferior in troops and materiel this force inflicted significant defeats on the advancing Italians, notably at Sidi Barrani, Bardia and Tobruk.

In August Pattle's No.80 squadron relieved No.33, also a Gladiator unit which had been holding the area since the outbreak of hostilities and on 4 August Pattle made his first victory claims with the downing of a Breda 65 bomber and Fiat CR32 fighter. During the combat Pattle and another pilot were forced to take to their parachutes, both coming down behind the Italian lines. They managed to make their way on foot to safety and reported to their squadron on the following evening.

Only four days later Pattle claimed two Fiat CR42 fighters, however these were his last Western Desert combats for in November 1940 his squadron was despatched to Greece.

In October 1940 Mussolini had made an unprovoked invasion of Greece and although vastly outnumbered and lacking tanks and modern aircraft the Greek army resisted the invaders stubbornly. The involvement of British forces was the result of an assurance by Winston Churchill to assist Greece and Romania in the event of an axis invasion. His long-term vision was for the opening of a second front against Germany via the Balkans but as events were to prove, such a scheme was quite beyond the resources of the Allies.

Meanwhile the Gladiators commenced operations from a forward airfield at Trikkala from where they began a series of offensive patrols over the front lines where some hard- fought combats took place. On 19 November Pattle claimed two Fiat CR 42s of the nine credited to the squadron and on 2

December he downed two RO37 reconnaissance biplanes. Despite the onset of winter and a reduction in the level of ground operations Pattle's squadron still remained active, mainly against bomber formations. This period also saw an even heavier air battle, involving over 50 escorting Fiat CR42 fighters, ten of which were brought down for the loss of two Gladiators. These combats were significant in the fact that they were some of the last encounters between biplane fighters in modern warfare.

At this point the harsh Greek winter clamped down, putting an end to operations for several weeks and in late February No.80 squadron began receiving Hurricanes to supersede their obsolescent Gladiators. This was particularly opportune, for the opposing Italian fighters now included monoplane types such as the Macchi 200 and Fiat G50. Meanwhile Pattle's score mounted steadily, with successes against these newcomers which represented his last combats with No.80. On 12 March Pattle, now a recipient of the DFC and Bar and with a confirmed victory tally of 23 was posted to command No.33 squadron, also a Hurricane unit.

On 6 April 1941 the anticipated German invasion of the Balkans began and with their overwhelming superiority in troops, armour and aircraft it was the beginning of the end for Allied forces in Greece. It was also the point that No.33 squadron records ceased to exist; however its remaining pilots fought desperately in a one-sided rearguard campaign. On 20 April Pattle led 15 Hurricanes to challenge over 90 Luftwaffe fighters and bombers in an operation remembered as the Battle of Athens. Pattle was observed as having despatched a BF109 and BF110 but in going to the rescue of a fellow pilot, Pattle himself fell to the guns of two BF110s who closed from behind and shot him down into the sea off Megara. He was quickly avenged by a Hurricane pilot who brought down both of Pattle's attackers but the sad fact remained that Pattle had fallen. During his tenure with No.33 he was credited unofficially with 30 victories and this combined with the 23 from No.80 places him as the top Allied ace but with the loss of No.33 squadron records this can never be substantiated.

Douglas Bader

W/Cdr. Douglas Bader (1910-1982)

Douglas Bader's career was the perfect example to mankind that life does not end when one is confronted by adversity. Cruelly injured in an air crash in December 1931 his flying days would appear to have ended. Instead by sheer determination he overcame the loss of both legs to eventually resume his interrupted RAF career and become a legend in his own lifetime. Born in London in 1910 Douglas Robert Steuart Bader was younger by 18 months than his brother Frederick. Bader Minor proved to be an energetic child, adventurous and one who excelled in robust sports, particularly rugby. The boys' father, originally a civil engineer, had enlisted in the Royal Engineers at the outbreak of war in 1914, suffering shrapnel head wounds in 1917, and although seemingly to have recovered he died of his wounds in 1922.

Douglas Bader's path to an aviation career would seem to have been initiated by the marriage of his aunt Hazel to Cyril Burge, who had flown in the RFC during the war and opted to remain in the post war RAF. His nephew's active imagination was fired by the exploits of those pioneer air fighters and on completion of his formal education he was determined to join the RAF. Fortunately the question of his mother finding the £150 annual fee for the two year course at the RAF College at Cranwell was solved by Douglas gaining a prize cadetship to the academy. It was September 1928 and he was eighteen years old.

Bader was taught to fly in the Avro 504 and in his assessment was judged as 'above average'. Off-duty periods were taken up by sport, where Bader represented Cranwell at rugby, cricket, hockey and boxing. In July 1930 he was commissioned as a fully trained pilot and posted to No.23 squadron at Kenley. This unit operated with the Gloster Gamecock, one of the last to use the agile little biplane. For the past two years 23 Squadron had been picked to perform the combined aerobatics at the RAF Hendon pageant. Bader's burning ambition was to be one of the two pilots in any future display, and in that he was successful, combining with his C.O. Harry Day in a scintillating exhibition at the 1931 air show.

Later in the year the squadron's Gamecocks were replaced by the Bristol

Bulldog, faster than the Gamecock but less agile in aerobatic mode. It was in December 1931 that Bader suffered a horrendous crash while performing unauthorised low-level aerobatics. This impulsive action robbed him of both legs and also it seemed a most promising flying career. After months in hospital, when at times he was near to death, he rallied and was fired with a determination to face life again. Fitted with duralumin artificial legs Bader made such extraordinary progress that by September 1932 he was able to fly again. Unfortunately there was no regulation to cover a legless pilot and to his dismay he was obliged to retire from the service.

During the 1930s Bader worked in a clerical position with the Asiatic Petroleum Company and in that somewhat barren era he found some relief from his frustration in playing golf and squash. He also married a remarkable young woman, Thelma Edwards who proved to be a staunch and understanding companion for the volatile Douglas. With the threat of war imminent Bader haunted the halls of RAF headquarters in the hope of a return to the service. His persistence was rewarded and in November 1939 Flying Officer Bader returned to active duty.

His first posting was to No.19 Spitfire squadron, however his first taste of combat occurred after his transfer to 222 squadron. This was in June 1940 at the time of 'Operation Dynamo', the retrieval of over 300,000 British and French troops from the heavily bombed beaches of Dunkirk.

With his obvious dash and ability Bader was given command of No.242 squadron, based at Coltishall near Norwich. This was a Hurricane unit, part of No12 Group and composed mostly of Canadians. It had been sent to France only weeks before the French capitulation and during its brief Continental tenure it had been badly mauled and lost quite a few aircraft and pilots. Not surprisingly morale was low and the squadron needed the impetus of strong leadership. Some personnel were at first amazed that their new C.O. was actually legless, but in leading by example Squadron Leader Bader rapidly forged a disgruntled squadron into an eager, cohesive unit.

August 8 was remembered as the official first day of the Battle of Britain and from late August to late September Bader and his Canadians were in almost continuous action. Despite the overwhelming odds they enjoyed tremendous success against these massed formations. One instance was the engagement of nine Hurricanes against 100-plus enemy aircraft, resulting in 12 victories and no losses to Bader's men. During the Battle No.242 claimed a total of 63 destroyed for the loss of three of its own, and of that total ten were gained by Bader himself; a period in which he was awarded the DSO and DFC. A significant aspect was Bader's 'Big Wing' theory, where a large fighter wing would intercept enemy formations rather than piece-meal sorties by single squadrons. With Air Vice Marshal Leigh Mallory's sanction it was Bader who led the wing, comprising 19, 242 and 310 squadrons and they did achieve worthwhile successes.

Following a successful outcome of the Battle, Fighter Command went on to the offensive by taking the air war over the Channel to the enemy. Early in 1941 Bader assumed command of the Tangmere Wing, taking Nos. 145, 610 and 616 into the skies over northern France. The intention was to draw the Luftwaffe into combat and in that respect the Germans were ever eager to respond. During June and July Bader was in the thick of the action and adding regularly to his tally, which by that stage had reached twenty. This period would be the pinnacle of Bader's impressive combat career; supremely confident and outwardly impervious to fear, his pilots regarded him as being indestructible as he led them on their 'Sweeps' and 'Circus' operations.

However on 11 August everything went wrong from the start; first there was confusion on take-off and the top-cover squadron went astray. It was over Lille that Bader's Wing tangled with a gaggle of a dozen BF109s and in the ensuing melee he found himself alone in a hostile sky, but then saw six 109s ahead of him. Unnoticed he slid in behind them and quickly despatched two of their number, but as he banked away his Spitfire was forcibly struck from behind. Looking back he was appalled to see that the entire fuselage behind the cockpit had been sheared off in a possible collision with another 109; (In recent years there is the suggestion that he was brought down by 'Friendly' fire; by a

British pilot in an unfortunate case of mistaken identity. Meanwhile his Spitfire was spiralling earthwards and in trying to bail out his right leg became firmly trapped in the cockpit.

Finally the leather and steel snapped and he was floating in space. In a flash his brain cleared and he pulled the D-ring, hearing a crack as the parachute opened. Bader floated to earth minus the offending leg and after a heavy landing in which he was temporarily knocked out he was taken prisoner.

For the next three and a half years Bader became a constant nuisance to his captors with various attempts to escape until finally he was sent to the notorious Colditz Castle, and in May 1945 its rebellious inmates were eventually liberated by advancing American troops. With the end of hostilities in Europe he was denied the opportunity to resume operational flying and in the official RAF list of fighter aces he was credited with 22½ victories, placing him 18th amongst illustrious company. Most pilots have their private victory total which in Douglas Bader's case was 30. It was a remarkable achievement for an individual who was obliged to retire from the service in 1932.

In a post-war RAF Bader rose to the rank of group captain in command of the North Weald Fighter Sector, and later in the year he organised and led the victory fly-past over London on 15 September, the anniversary of the greatest day of the Battle of Britain. He retired from the RAF in July 1946, returning to Shell Oil where he became a senior executive. Douglas Bader was later knighted for his courage and determination and tireless work for the handicapped, which continued to inspire future generations. He died in 1982; that doughty heart which sustained him throughout a tempestuous life finally cried enough.

Clive Caldwell

W/Cdr. Clive Caldwell (1910-1994)

Clive Robertson Caldwell gained fame as the top-scoring Australian fighter pilot of World War 2. Tall and athletic, he was an excellent marksman, a brilliant pilot and potential leader of men. In a hectic career that spanned four years of combat he flew Tomahawks and Kittyhawks in North Africa and Spitfires in Europe and the South West Pacific. During that period he was in the unique position of having fought the Germans, Italians, Vichy French and the Japanese and shot down at least 30 enemy aircraft. The future ace was born in Sydney in 1910 when the aeroplanes that were to play a large part in Caldwell's career scarcely existed. It is worth noting that twelve days prior to his birth a young Australian, John Duigan had built and flown the first locally-designed aircraft at Spring Plains in Victoria. Three years would elapse before Australia set up its own air force with the formation of the Australian Flying Corps at Point Cook in Victoria in 1913.

Caldwell was born into a middle-class Sydney family and was educated at Trinity Grammar where his academic career was generally unremarkable, although he excelled in athletic pursuits and boxing as a teenager. His first employment situation was with the Bank of New South Wales but the restless Caldwell rebelled against the rigidity of the banking system and found other employment in a variety of jobs. One of these was as a jackaroo at an outback sheep station where in off-duty periods he was able to hone his shooting skills, to the detriment of a future generation of opposing airmen.

In 1938 Caldwell's finances were sufficient for him to take flying lessons at the Royal Aero Club of N.S.W. After only three and a half hours of dual instruction he made his first solo and with the approach of war he had accumulated time on Tiger Moths. With the outbreak of war in September 1939 Caldwell was eager to be a fighter pilot, however the cut-off point for single-seater training was 28 and only by altering the details on his birth certificate did he manage to be accepted into the RAAF. To his dismay he discovered that those on his course were to be trained as instructors, a situation which held no appeal to Caldwell. He promptly discharged himself and in April 1940 for a second time he joined the RAAF, this time as a trainee in the

Empire Air Training Scheme. His competence as a pilot was obvious and on gaining his wings he was impatient to be sent to England to fight in the Battle of Britain, however by October it was all but over. Instead he embarked for the Middle East in February 1941, where as Pilot Officer Caldwell he joined 250 squadron RAF based at Aquir in Palestine.

This unit operated with the Curtiss P40C Tomahawk which had performed valiantly as a front-line fighter with the US Army Air Corps during 1941 and 1942, although in RAF service it was rejected as a fighter in Western Europe. Their Allison engines were rated to only 12,000 feet, rendering combat above 15,000 feet totally impractical; fortunately for the P40 pilots in their encounters with the BF109 and Macchi 200 the desert war was generally a low-level affair. Initially, Caldwell like so many other tyro pilots had little success in his early sorties, until his 33rd mission when he scored his first confirmed kill; a BF109 plus a share in the destruction of a Cant Z1007 bomber near Alexandria.

Caldwell's reputation as an air fighter became legendary, as witnessed by the mounting number of crosses stencilled on his P40. In January 1942, with his victory total at 18 Caldwell was promoted to squadron leader and given command of 112 squadron RAF. This was a commendable achievement, being the first Empire Air trainee to command a squadron and with the prestige of the award of the DFC and Bar. This unit had recently converted to the improved version of the Tomahawk, the P40E Kittyhawk. Although inferior to the BF109 they performed well in a variety of roles, particularly in bombing and strafing of enemy transport and airfields.

Caldwell's tenure with 112 ended in May 1942 when he was ordered to report to England. At that point his victory tally was 20 kills after 550 hours of operations. His passage to the United Kingdom was performed over a series of flights via West Africa, The Caribbean, The United States and Canada, finally arriving in Scotland in late May 1942. By June he was back on operations commanding the Kenley Spitfire Wing, and in a period of ground attacks over France he added a locomotive to his list of kills.

With Japan having entered the war Australia was experiencing air raids on a regular basis on Darwin and other Northern Territory towns. As a result Caldwell was recalled to Australia in November 1942 as wing leader of the newly-raised No.1 Fighter Wing. This unit which comprised former UK-based Spitfire squadrons No.452 RAAF, No.457 RAAF and No.54 RAF arrived in Darwin in January 1943. Their aircraft were Spitfire Vcs fitted with tropical filters to cope with local conditions. The fighter wing began operations over northern Australia with its focus on the defence of Darwin which at the time was a major factor in the defence of Australia.

On 2 March 1943 Caldwell claimed his first Japanese victories while leading a flight of six Spitfires. His formation intercepted 6 Kate dive bombers escorted by 12 Zero fighters about to attack Allied shipping in the Arafura Sea north of Darwin. In the ensuing melee Caldwell destroyed one Zero and one Kate and by August Caldwell had despatched 8 Japanese aircraft, elevating his tally 30.5. He was then taken off operations and posted to No.2 Operational Training Unit as Chief Flying Instructor.

In May 1944 Caldwell returned to operations as wing leader of No.80 Fighter Wing based in Morotai. This unit which comprised No.79, No.452 and No.457 squadrons operated with the far-superior Spitfire Mk.VIII and rather than the fighter role it was employed in ground attack sorties. In fact their Spitfires were ill-suited to these duties, proving quite vulnerable to ground fire in what the pilots regarded as unimportant targets.

By the end of 1944 it was apparent they were being left out of the main arena of operations and as a result of nothing having been done to meet the pilots' demands Caldwell and two other senior officers resigned in protest. This affair gained notoriety as the 'Morotai Mutiny' and led to a command crisis in the RAAF. An investigation resulted in two senior officers being relieved of their appointments. Caldwell was eventually reinstated and finished the war attached to HQ Ist Tactical Air Force RAAF. He resigned from the RAAF in 1946.

In a post-war situation Caldwell created a wholesale fabrics business. Failing health caused him to relinquish his management of the enterprise and on 5 August 1994 Australia's top-scoring ace passed away.

Pierre Clostermann

W/Cdr. Pierre Clostermann (1921-2006)

Pierre Clostermann heads the list of French aces, credited with 23 official victories. These were achieved over a remarkable 432 sorties while serving with 341 'Alsace' squadron and various RAF units between January 1943 and the end of hostilities in Europe in May 1945. Born in Brazil in 1921, the only child of a French diplomatic family the young Clostermann completed his secondary education in France. At age 16 he gained his private pilot's licence and on the outbreak of war in September 1939 he applied for service in the Armee de l'Air, but to his dismay this was refused. As a result he sailed to a then-neutral America to become a commercial pilot, studying at the California Institute of Technology.

In March 1942 Clostermann joined the Free French Air Force in Britain, and following training at RAF Cranwell he was assigned to an operational training unit in Wales for a two-month course on Spitfires. On completion, with the rank of sergeant pilot, Clostermann was posted to 341 squadron RAF based at Turnhouse, Edinburgh. This new unit gained fame as the 'Alsace' squadron, composed of Free French pilots who journeyed from far and wide to be part of this unique force. In a surprisingly short time the Alsace squadron was melded into an efficient unit and one month later was posted to the Biggin Hill Wing, south of London. It was a singular honour for so new a squadron to be selected, for prestigious Biggin Hill was the base with the highest number of victories to its credit. At this stage of the war Fighter Command was involved in mounting large-scale offensive sweeps over northern France. To counter these actions the Luftwaffe reacted strongly, notably with the Focke-Wulf 190, which was appearing in increasing numbers. Front-line RAF squadrons were being equipped with the Spitfire IX, at the time the last word in aero- technique and fairly evenly matched against the FW 190. It was during Clostermann's second such mission that he scored his first combat victories with the downing of two FW 190s.

In September 1943 Clostermann was posted to 602 ('City of Glasgow') squadron based at Ashford. During the Battle of Britain this unit had played a starring role and since then had been relegated to a secondary position. It was one of the first units to be transferred to the Tactical Air Force in preparation for the eventual invasion of the continent when it would provide close co-operation with the Army. After the glamorous status of Biggin Hill, 125 Airfield gave a rather 'country cousin' impression and for four months the pilots lived under canvas learning to refuel re-arm and camouflage their aircraft and also defend them, leading a real 'commando' existence. For the moment they operated with the Spitfire V-D, a clipped-wing variant which gave improved manoeuvrability and speed at low altitudes, but with its lower rated Merlin 45 engine the performance fell away markedly above 5000 feet. Nevertheless the squadron was required to carry out further cross-Channel sweeps before the arrival of their brand new Spitfire IXs, In the meantime it flew fighter sweeps, bomber escorts and dive-bombing and strafing attacks on V-1 launch sites on the French coast.

In January 1944 the squadron was ordered north to Skeabrae in the Orkneys, which in a northern winter approximated conditions akin to the North Pole. Their role was to foil any attempts at bombing or reconnaissance on the part of the Luftwaffe of Royal Navy units at Scapa Flow.

Apart from their usual Spitfire Vs there were four special Strato-Spit VIIs available, each capable of operating up to 45,000 feet. Their opportunity for combat came in February when Clostermann and another pilot intercepted and destroyed a BF 109 on a photo-reconnaissance mission at the remarkable altitude of 41,000 feet. Shortly afterwards the squadron returned to southern England in preparation for the imminent invasion of Europe. It was a period of intense aerial activity involving fighter sweeps and dive-bombing of targets in France. On D-Day itself, 6 June the squadron carried out three sweeps over the Normandy area on a momentous occasion for the Allied forces involved. On II June the squadron was mobilised to spend the night in France from an improvised landing ground, which nevertheless was an emotional moment for an expatriate Frenchman to be treading French soil again. However their sojourn was relatively brief for the squadron returned to England on the following day. However on 17 June the unit returned to France where it operated from a temporary airfield (B11) near Arromanches. This was a period of hectic combat against a most aggressive Luftwaffe when Clostermann added steadily to his victory tally, but 16 months of continuous operational flying had taken its toll and on medical advice he was recalled from operations and transferred to an administrative position with Free French H.Q. in England.

In December 1944 Clostermann made the decision to return to active operations, and despite a blunt refusal by the Paris-based Ministere De I'Air to permit this, it was through the influence of Colonel Coustey, O.C of French Air Forces in Great Britain that permission was eventually granted. Following a conversion course on Typhoons and Tempests Clostermann returned to operational status on secondment to No274 Tempest squadron as a supernumerary flight lieutenant in command of A flight. This unit was part of 122 Wing and for the moment was based at Volkel, Holland from where it was in constant action against a determined Luftwaffe equipped with the latest FW190s and BF109s and also an increasing number of a new generation of ME262 jet fighters. In his personal Tempest 'Le Grand Charles' Clostermann flew an intensive round of fighter sweeps, airfield attacks and rail interdiction missions across northern Germany over the ensuing two months.

In March 1945 Clostermann served briefly with another Tempest unit, No.56, prior to a transfer as wing commander of A flight with No.3 squadron. Following the German surrender on 5 May the Tempest Wing took part in a victory fly-past at Bremerhaven, and with it a tragedy as the

Tempests in Clostermann's section became catastrophically tangled up at less than a thousand feet. For the first time in his hectic career Clostermann was forced to bail out and despite the minimum altitude he survived; unlike three of his comrades who perished in their attempts. In July 1945 Clostermann was demobilised from the RAF and in a post war era he achieved fame in a variety of pursuits, notably as an author with two best-selling books; 'The Big Show' and 'Flames in the Sky'. He also served eight terms as an MP in the French National Assembly. He died in March 2006 and in a tribute to his return to France in 1944 a road in Longues-sur-Mer was named after him; this occurred in June 2004, sixty years after that memorable event.

Richard Bong

Major Richard Bong Top-scoring U.S. Ace

Richard Ira Bong who was destined to became America's 'Ace of Aces' of World War 2 was born in 1920, the son of a Swedish immigrant father and Anglo-Scots mother. His early years were spent on the family farm at Poplar Wisconsin. In essence he was the typical American country youngster, working on the family farm and gaining high grades at school. He was also a keen sportsman, playing on the school's baseball, basketball and hockey teams. Fishing and hunting were abiding interests for the teenager at which he became proficient with a hunting rifle. Like many of his contemporaries Bong developed an early interest in aviation and during his secondary education at Superior State Teachers College in 1938 he enrolled in the civilian pilot training programme and also took flying lessons. In early 1941 with America yet uncommitted to the European conflict Bong enlisted in the Army Air Corps aviation cadet programme.

Bong's primary training was conducted at Rankin Aeronautical Academy in California in June 1941, followed by basic training at Gardner Field, California. From there he was sent to Luke Field Arizona for advanced training in single-engine (fighter) aircraft. His machine was the rugged AT-6 Texan (or Harvard in RAF service) which launched the bulk of trainee Allied pilots on their chosen career. Bong proved to be a natural pilot, obviously suited to fly fighters and in January 1942, just after Pearl Harbour he gained his Army Air Corps commission and those coveted pilot's wings. But instead of an anticipated career as a fighter pilot Bong was retained as a gunnery instructor.

This tenure lasted several months when his opportunity for action was presented by the introduction of the Lockheed P38 into Air Corps service. Bong quickly adapted to the challenge of the somewhat-intimidating twin-engine Lockheed, however his low-flying escapades around San Francisco bay soon attracted the attention of General George Kenney, head of the Fifth Air Force. No doubt the budding ace would have incurred the general's wrath for his undisciplined aerobatics but significantly Bong was one of the pilots selected to join the 9th Fighter Squadron. He was shipped to Australia pending the delivery of the unit's P38s, meantime Bong spent time with the 35th Fighter Group operating out of Port Moresby New Guinea. During his tenure with the 35th he opened his account with the destruction of two Japanese fighters, a Zero and an Oscar.

During the Pacific conflict Japan produced no fewer than 118 different types of military aircraft and in an effort to simplify identification American Air Intelligence allotted a distinctive name to each type. In early 1942 all enemy fighters were called Zero or Mitsubishi if they had more than one engine. However this was a dangerous over-simplification when evaluating the potential strength of one's enemy. Eventually each aircraft was given an arbitrary title; with fighters and floatplanes receiving male Christian names; e.g. *Zeke, Oscar, Rufe* etc.; with bombers and flying boats given female Christian names; e.g. *Betty, Sally, Mavis* etc. It proved to be an efficient arrangement and from it there emerged a host of classic types, many of which proved equal in fame to the best of the Allied opposition.

Meantime during 1943 Bong's score was steadily rising, with each month proving successful. Mostly these were individual figures except on 26 July when he claimed four fighters while on bomber escort over Lae in northern New Guinea. By December 1943 Bong's tally had reached 21, at which point he was sent on home leave to Wisconsin. During this welcome break from operational flying he met Marge Vattendahl, a young woman whose name and portrait would be featured on Bong's P38 and whom he would marry in February 1945.

In February 1944 Bong returned to the New Guinea theatre where he resumed operations, this time in the 'free-lance' role, a situation that suited his individualistic nature. By April 1944 he had surpassed Eddie Rickenbacker's WW1 record of 26, thus making Bong the current top-scoring American of two world wars. At this point General Kenney promoted Bong to major and for a second time he returned home to The States where during May-July he made a series of publicity tours. These were not to Bong's liking and in September he was re-assigned to the Southwest Pacific area as an advanced gunnery instructor at Hollandia. General Kenney did allow Bong to go on missions but not to seek combat and merely defend himself whenever necessary. Nevertheless by early October he added three more to his victory tally and with the American landings in the Philippines Bong successfully lobbied to be included in this crucial action. By mid-December 1944 Bong increased his score to forty and at that point he was finally grounded by General Kenney and sent home. Prior to that Bong was presented with the Congressional Medal of Honour by General MacArthur on Tacloban airfield.

By New Year's Eve the 'Ace of Aces' was back in The States where he was sent on a PR tour to stimulate public interest in purchasing war bonds. That duty completed, in February 1945 he married Marge and after a Californian honeymoon he began training as a test pilot at Wright Field in the development of the new Lockheed P80 Shooting Star. This involved a two-month theory course on jet propulsion before getting the chance to fly one. Finally Bong was checked out on the P80 and during that summer flew it on eleven occasions. August 6 1945 is a date indelibly remembered as the day on which the first atomic bomb was detonated over Hiroshima. It is also significant as the occasion of Richard Bong's final flight. After defying the odds on 200 combat missions over 500 combat hours he became a victim of the law of averages when his P80 suffered a flame- out on take-off. Bong made an emergency bail-out but with insufficient altitude he perished in the attempt. This was a sad conclusion to a brilliant aviation career and especially tragic to be betrayed by mechanical failure. During Bong's operational career he described combat flying as great fun and not being overly concerned with victory scores but rather the zest of actual combat. This attitude differed greatly from other aces whose sole concern was their scores, almost to the point of recklessness. Bong on the other hand felt no shame in abandoning combat when the situation was unfavourable. His closest rival on the closing stages of a remarkable operational career was another P38 exponent, the great Tommy McGuire with 38 scalps to his credit and the acknowledged Pacific ace following Bong's departure.

McGuire however was not destined to overtake Bong's tally of 40 when he too became a victim of the law of averages. On 7 January 1945 he led a flight of experienced P38 pilots, Rittmayer, Thropp and Weaver on an early morning fighter sweep over the Japanese airfield at Negros Island. However inclement weather forced them to proceed to another airfield on the western side of the airfield. Abruptly a lone KI-43 *Oscar* flew through the group, passing below McGuire and his wingman before either pilot could react. Meanwhile a KI-84 *Frank* joined the melee and inflicted damage on Thropp's P38; while Rittmayer who had earlier experienced engine trouble managed to deter the *Frank* off Thropp's tail. Rather than retreat the Japanese pilot chose to stay and harass the P38s, firing bursts into Weaver's aircraft. At that point Weaver summoned McGuire's assistance, whose response was immediate but as he banked steeply to his left his Lightning shuddered, lost momentum and was last seen inverted and nose down, followed seconds later by an explosion. Tommy McGuire had fallen and the aggressive pilot of the *Frank* inflicted further loss to the Americans when Major Rittmayer was brought down near the area where McGuire had crashed a few minutes earlier. The surviving P38s, both suffering damage managed to escape the situation and return to base. It had been a tragic outcome, particularly so for McGuire being denied the opportunity to overtake Bong's tally and ironic that the Congressional Medal of Honour award he so coveted would be a posthumous one.

Chuck Yeager

Col. Charles Yeager Breaker of the Sound Barrier

In the course of an aviation career spanning 60 years Charles Elwood (Chuck) Yeager emerges as a truly remarkable airman and unquestionably the most famous test pilot of all time. Even his surname 'Yeager' was prophetic, being an Anglicised version of the German 'Jager' or 'Jaeger' (Hunter) He was born 13 February 1923 to farming parents, Susie and Albert Yeager in Myra West Virginia and graduated from high school in Hamlin West Virginia.

His initial military experience was at a Citizen's Military Training camp at Fort Benjamin Harrison, Indiana. This was during the summers of 1939 and 1940, and prior to America's entry into World War 2 in 1941 Yeager enlisted as a private in the USAAC. Following service as an aircraft mechanic he was accepted for pilot training; gaining his wings and promotion to Flight Officer at Luke Field Arizona. Assigned to the 357th Fighter Group at Tonopah Nevada, Yeager initially trained on the Bell P39 Airacobra, an unorthodox type with its engine mounted behind the pilot.

In November 1943 the Group was shipped overseas to the United Kingdom, stationed at RAF Leiston. Flying the P51 Mustang with the 363rd Fighter Group in combat over Occupied Europe, Yeager claimed one victory before he was shot down over France on his eighth mission. Managing to evade capture Yeager escaped over the Pyrenees to neutral Spain with the help of the French Resistance (The Maquis). During his time with the Maquis, Yeager assisted them with various duties, involving bomb construction and other non-direct combat tasks.

He was subsequently repatriated back to the United Kingdom and to his dismay he was forbidden to fly over enemy territory again to avoid compromising their Resistance allies. Yeager pleaded his case to General Eisenhower on 12 June 1944, pointing out that the Allies had already invaded France and the Maquis was by then openly fighting the Nazis. Back with his fighter squadron Yeager distinguished himself by becoming the first pilot in the Group to claim five enemy aircraft in one mission.

He finished the war with 11.5 victories, including one of the first air-to-air kills over a jet fighter, a German Me262. Yeager's Mustang was named 'Glamorous Glennis' after his girl-friend Glennis Dickhouse, whom he married in February 1945; it was a tradition he maintained on his future aircraft, including the iconic Bell X-1 rocket plane, in which he recorded record-breaking flights in 1947.

He flew his 61st and final mission on 15 January 1945, returning to The United States in early February. Yeager chose to remain in the Air Force in the post-war era, and with his distinguished fighter pilot record, along with being involved in aircraft maintenance prior to becoming a pilot, he was an obvious candidate for the test pilot role.

Yeager became part of the Air Force team at Muroc Army Air Field; now Edwards Air Force Base, located in California's Antelope Valley. In its favour was the generally stable weather pattern and clear unlimited visibility; so vastly different to the United Kingdom and European conditions.

Supersonic flight was still a challenging exercise in those early post-war years, with many fine pilots, including Geoffrey de Havilland making the supreme sacrifice in their quest to conquer the so-called 'sound barrier'.

The Bell Aircraft Corporation was renowned for its range of innovative fighter aircraft, and their experimental X-1 rocket plane was no exception. Beautifully streamlined, the rocket- powered X-1 was designed to be taken up to altitudes of 45,000 feet by a 'mother' plane, in this instance a B29 Superfortress.

Two days before the scheduled flight Yeager had the misfortune to fracture two ribs following a fall from a horse. Rather than disclose the injury, with its excruciating pain, he carried on with the test flight, only revealing the details to his wife, one other and fellow project pilot Jack Ridley. It speaks volumes for Yeager's tenacity that the record attempt was successful, with the X-1 recording Mach 1.07 on 14 October 1947.

In 1948 Yeager was awarded the McKay and Collier trophies for his mach-transcending flight and in 1952 he created a new air speed record of 1650 mph, more than twice the speed of sound. During the Korean conflict Yeager became one of the first American pilots to fly a captured Mig15 after its pilot defected from North Korea. He later commanded a fighter squadron in Europe at the height of the 'Cold War'.

In 1956 Yeager assumed command of the Air Force Aerospace Research Pilot's School for the emerging space program, and although he was passed over for service in space, a high percentage of the new generation of astronauts were graduates of Yeager's school. In 1966 NASA took over the training of astronauts, resulting in the closure of the Air Force School. During the Vietnam War Yeager, now a full colonel, commanded the 405th Fighter Wing out of the Philippines, flying 127 air-support missions and training bomber pilots.

In 1968 Yeager was promoted to brigadier general, making him one of the very few to have risen from enlisted man to general in the Air Force. Among the many awards conferred on this remarkable airman, perhaps the most prestigious were the Congressional Gold Medal presented to him by President Gerald Ford and the Presidential Medal of Freedom from Ronald Reagan.

Yeager retired from the Air Force in 1975 but continued to serve as a consulting test pilot for many years. October 14 1997 was the 50th anniversary of the history-making flight in the Bell X-1 and in a significant gesture Yeager observed the occasion by breaking the sound barrier again, this time in an F-15 fighter.

Author appreciation

This concludes an insight into a century of man's quest to master the astral elements. Orville and Wilbur Wright's experiments virtually paralleled the horse-and- buggy era, while Chuck Yeager's exploits belonged to a vastly different universe. A sobering aspect of aviation progress was the intervention of two world wars.

This was forcibly demonstrated with the onset of World War 1 where the rickety types that bravely staggered into Western Front skies were developed into such iconic types as the Fokker D7 and Sopwith Snipe. It also saw the beginnings of aerial bombardment, as demonstrated by the multi-engine Handley Page O/100 and Vickers Vimy, versatile designs that were amenable to conversion as air liners in more peaceful skies.

After twenty years of uneasy peace and rising international tensions, once more the nations were plunged into another world war. In this holocaust aviation advancement proceeded at a relentless pace, in particular with the advent of the gas turbine. In less than a decade the piston-engine fighter had been relegated to the pages of history; so too in air travel with Britain's DH Comet of the early 1950s, which initially led the way to a new era in commercial aviation.

Perhaps the ultimate dream that became a reality was the supersonic Concorde of recent memory. One wonders what Wilbur and Orville Wright would have thought of Concorde? At least Orville was given an insight into the future when he became a visitor to the flight deck of the new Lockheed Constellation shortly before his death in January 1948. It was reported that he took the controls briefly; surely this was the most satisfying moment for the pioneer of Kitty Hawk?

Author Profile

Australian artist Murray McLeod has an abiding interest in historic aviation which he transposes into book form with a range of current titles. Another interest is his motorcycle racing titles, also available in book and Kindle mode. Murray welcomes enquiries for that *special* art work.

Author web site: www.mcleodart.com.au

www.ingramcontent.com/pod-product-compliance
Lightning Source LLC
Chambersburg PA
CBHW051151220526
45473CB00003B/737